The Cinematic Life of the Gene

[The Cinematic Life of the Gene]

JACKIE STACEY

DUKE UNIVERSITY PRESS   DURHAM AND LONDON   2010

© 2010 Duke University Press

All rights reserved

Printed in the United States of America

on acid-free paper ∞

Designed by C. H. Westmoreland

Typeset in Warnock by Achorn International, Inc.

Library of Congress Cataloging-in-Publication Data

and republication acknowledgments

appear on the last printed page of this book.

FOR ANNA

Contents

Preface ix

Acknowledgments xv

Introduction: Technologies of Imitation and
the Genetic Imaginary 1

[PART **1**] Sameness Ad Infinitum

1. The Hell of the Same: Cloning, Baudrillard, and the
Queering of Biology 19

2. She Is Not Herself: The Deviant Relations of
Alien: Resurrection 36

3. Screening the Gene: Femininity as Code in *Species* 66

[PART **2**] Imitations of Life

4. Cloning as Biomimicry 95

5. Genetic Impersonation and the Improvisation
of Kinship: *Gattaca*'s Queer Visions 113

6. The Uncanny Architectures of Intimacy
in *Code 46* 137

[PART **3**] Stairway to Heaven
7. Cut-and-Paste Bodies: The Shock of
Genetic Simulation 179
8. Leading Across the In-Between:
Transductive Cinema in *Teknolust* 195
9. Enacting the Gene: The Animation of Science
in *Genetic Admiration* 225
Afterword: Double Take, Déjà Vu 257

Notes 273
Bibliography 287
Filmography 303
Index 307

Preface

The Cinematic Life of the Gene is a cultural study of film that brings
the dynamics of visibility, embodiment, and artificiality center stage.
It explores the changing relationship between biological and cultural
forms at the current conjuncture of science, feminism, and the cinema.
Taking both films and theoretical texts as its focus, the book seeks to
contribute to the current interdisciplinary debate about the geneticization of the body, which stretches from cultural studies of science and
technology across feminist film criticism to queer theories of desire,
embodiment, and kinship.

Each of the three sections of the book contains one short theoretical
speculation, followed by two longer film readings. These sections are
organized around very different genres, united by a common concern
with the reconfiguration of the surface-depth relation in the production of a sense of geneticized embodiment. The readings that follow
focus primarily on six films released since 1995, in the decade in which
Dolly the sheep was the first mammal to be cloned from an adult cell.
All of the films explore the problem of making visible the sameness of
the identical biological copy. The film readings move from Hollywood

body-horror films, *Alien: Resurrection* (Jean-Pierre Jeunet, 1997) and *Species* (Roger Donaldson, 1995), in part I through the "intelligent" thrillers of "mainstreamed arthouse sensibility" (M. Hills 2005, 16), *Gattaca* (Andrew Niccol, 1997) and *Code 46* (Michael Winterbottom, 2003), in part II to comic independent work by feminist directors, *Teknolust* (Lynn Hershman-Leeson, 2002) and *Genetic Admiration* (Frances Leeming, 2005), in part III.

Beginning with the terror of the hidden depths of genetic interference, the genre of body horror unites the theoretical and cinematic focus in part I, "Sameness Ad Infinitum." In all three chapters in this first section, the full affective force of the genre is mobilized to push the potential new disturbance of genetically engineered biologies to its limit (although, as I discuss, repetitions between cinematic and philosophical generic repetitions introduce a self-conscious play on formal doubling). In chapter 1, a close reading of Baudrillard's writing about cloning (1993, 1997, 2000, and 2002) takes his apocalyptic proclamations about the prospect of the "hell of the same" as indicative of the logic of a cultural imaginary troubled by the convergence of genetic engineering with the queering of kinship and the uncoupling of sexuality from reproduction. The reading of *Alien: Resurrection* that follows in chapter 2 explores the associations of excessive sameness across the boundaries of disease, sexuality, and technology. Chapter 3 then examines the mutating boundaries between cultural and biological forms of disguise in *Species*. Both these chapters discuss how the films display the spectacular monstrosities that genetically engineered hybrid female reproductive systems body forth. Each film is read in terms of how genetic interventions reconfigure sexual and reproductive drives, how the autogenerative bodies of the female protagonists are the sites of horrific recombinant monstrous births, and how the smooth surfaces of idealized white feminine perfection erupt to reveal the inner disturbances of genetic indeterminacy. In this section of the book, feminist and queer readings of these films consider the heteronormative logic of a genetic imaginary that places sexual difference as the required foundation not only of desire itself, but also of all desirable forms of cultural and biological reproduction.

A shift to the genre of the art-house thriller takes us away from the spectacle of cellular body horror and into the clinical mise-en-scène of the pleasures and dangers of artifice and deception in part II, "Imitations of Life." Moving from the monstrous feminine to the *femme fatale*, this section concerns the genre's preoccupation with the danger of misreading clues and the problem of how to read identity in a world where artificiality plays with the illegibility of surfaces as indicative of hidden genetic coding. Here, surface-depth relations are never what they appear to be, and the gene as code translates into spatial deception as the diegetic promise of transparency eludes the desires of the detective. Opening with a short discussion of theories of imitation and bridging feminist readings of masquerade, postcolonial conceptualizations of mimicry, and queer notions of impersonation, chapter 4 looks backward to the visual camouflage of monstrous duplicity of the femme fatale in *Species*, and forward to the deliberate impersonation of genetic perfection in *Gattaca*. Chapter 5 then traces the cinematic language of the gene through the stylization of the techniques of genetic impersonation in *Gattaca*. Extending this play to the overlapping desires for artifice on the cinema screen and in the laboratory, *Code 46* is the subject of chapter 6. Here the architectures of globalized mobility and cosmopolitanized cities provide the mise-en-scène for a reimagined biological body: put simply, in these films, the geographical spatializes the biological. In a world of multicultural flow and monocultural stasis, biological purity is visualized through a genetic lens in which incestuous transgressions become Oedipal problems. Like *Gattaca*, *Code 46* rehearses the desire for technology to deliver transparency and security, while staging a world of distorted reflections, confusing appearances, and identity fraud. Biogenetic manipulation here is embedded within fantasies of high-tech regulation that tests the limits of these "control societies," where every move is monitored (Deleuze 1992).

In both films, the imitative techniques of genetic engineering are tested by the ingenuities of impersonation motivated by human desires. These two science-fiction thrillers are tied together through their stylish art-house aesthetic, which fetishistically displays the visual pleasures of one technology (cinema) to produce a critique of another (cloning). Bringing together theories of the masquerade in film and considerations of genetic techniques of impersonation, this section of the book maps out the ways in which sexualized and racialized bodies trouble the desire for technology to guarantee biological purity and genealogical truths. Just as the theories of mimicry explore the artifice of identity production, so the thriller genre turns biological codes into enigma and mystery.

Part III, "Stairway to Heaven," moves beyond the problem of the guarantee of visual evidence that may deceive us with surface impressions to consider the genetic imaginary as a disturbance to our modes of perception. Beginning in chapter 7 with a discussion of the geneticized body in light of Benjamin's (1968 [1936]) theory of the art object's loss of aura in the age of mechanical reproduction, this section explores the surface-depth tropes of phantasmatic projections. Here, both the reproducibility of the art object through mass production and circulation and the replication of the biological unit of information (such as DNA) through genetic engineering raise questions about the loss of a sense of originality and authenticity. Rather than focusing on the surface disguises at stake in the previous section, here recombinant biological forms are placed in dialogue with changing genres of mass culture. From the cautionary tales of the generic conventions governing cinematic and scientific masquerade in part II, we now shift to feminist independent films that take popular technological form itself as the object of inquiry. The affective impact of such formal disturbances to traditional relations between humans and artifacts is read through the question of changing modes of perception.

Considering the analogies between the image and the body in the "culture of the copy" (Schwartz 1996), chapter 8 pushes beyond notions of the interface into the domain of posthuman media convergence, "biomedia" (Thacker 2004), and "transduction" (Mackenzie 2002). In this reading of the blending of biogenetic and cinematic technologies in *Teknolust*, by the independent artist and filmmaker Lynn Hershman-Leeson, the chapter explores the comic mise-en-scène of artistic and filmic citation through which the film performs the techniques of imitation it narrativizes. The final film chapter of the book then looks at how *Genetic Admiration* uses collage animation techniques to undo our sense of awe at such biological and cultural convergences. It discusses how the film's transformative collage is animated in dialogue with our associative perceptions through a generative aesthetic that keeps both sound and image in motion, rehearsing familiar associations and then undercutting them with unconventional moves. Both *Teknolust* and *Genetic Admiration* use formal experimentation (pastiche and collage animation, respectively) to explore the restructuring of the body through genetic engineering and cloning, and to place the accompanying emergent fantasies within a repertoire of image cultures that unite artistic and scientific fields. Playing with generic formulas in very different ways, the two films discussed in this section rework motifs from the history of both popular science and popular cinema. Disrupting the flow from seeing to knowing to predicting, which forms the underlying desires of genetic engineering and cloning, each film undoes the associative logic of such teleologies and mocks the masculine vanity of such visions. Closing with two films by independent feminist directors—the one a citational pastiche, the other a deconstructive enactment—this part of the book examines the aesthetic promise of reassemblage, which turns the manipulation of bodies into an admirable art.

Each part of the book places films and theories in rather different relationships. In the first, Baudrillard's work is positioned as a symptomatic indicator of the condensations operating in the genetic imaginary. The genre of body horror describes both the film readings that follow and Baudrillard's theoretical interventions, in which philosophy becomes an eloquent science fiction that is as suggestive of the genetic imaginary as the iconographies of transgenic maternal monsters in the films. In the second part, detection becomes the trope for an indicative investigation of long-standing preoccupations with mistaken identity in both psychoanalysis and the cinema. The problem of misrecognition becomes newly biological, as genetic identity theft leads us from investigating crimes against the state into detecting crimes against nature. Just as the theoretical speculations in parts 1 and 2 might be read respectively as symptomatic and indicative, so the equivalent chapter in part 3 functions more analogically to consider the perceptual disturbance of genetic interference as a loss of bio-aura. Conventional detection becomes absurdly redundant here, when the status of the visual image as already imitative mocks our desire for it to act as a place of certainty and undercuts the push to unite the visual field with intelligibility, moving us out of conventional genres and into comic mimesis.

This three-stage journey takes the reader from the full force of the abject imaginary of genetically engineered and cloned futures in part 1 out into the suspense of deceptive surface appearances that threaten the possibility of not knowing the difference between original and unoriginal in part 2, and finally into the knowing spaces of artifice that play the power of scientific desires back to itself through the lens of feminism in part 3. But the ascent from full immersion to surface oxygenation is not indicative of a narrative of straightforward progress from popular cinema to the playful deconstructionist strategies of independent film. Running across this traditional divide are both the claustrophobic saturation of the inescapable dominance of geneticization and the breathing spaces of the inevitable generic repetitions of representational form.

Acknowledgments

This book was written in dialogue with colleagues in the Departments of Sociology, Women's Studies, and Cultural Studies at Lancaster University and at a number of other institutions, particularly the Center for Women's Studies at the University of Utrecht, the Center for the Study of Women and Gender at the University of British Columbia, and the Department of Communication Studies at Simon Fraser University. I am extremely grateful to Lancaster University for the periods of research leave which enabled me to complete this work, and to the Arts and Humanities Research Council (AHRC) for a matching leave research grant in 2005–6. During 2006 I spent two terms at the Center for Research in Women's Studies and Gender Relations at the University of British Columbia and enjoyed the warm hospitality and intellectual stimulation of its staff, visitors, and postgraduates. I am grateful to Sneja Gunew for her hard work in making this possible, and to Kirsten McAllister at Simon Fraser University for her facilitation of this period of leave in Vancouver. In 2007 I moved to the University of Manchester, and my thanks go to my new colleagues at the Research Institute for Cosmopolitan Cultures (RICC), at the Centre for the Study of Sexuality and Culture (cssc), and in English and American Studies for their generous welcome and for their support during the completion of this book.

I would like to thank the following people for their generous comments on earlier versions of chapters in this book: Marie-Luise Angerer, Lauren Berlant, Rosi Braidotti, Susan Brook, Lisa Cartwright, Claudia Castañeda, Richard Cavell, Julie Crawford, Peter Dickinson, Laura Doan, Zoë Druick, Richard Dyer, Anne-Marie Fortier, Sarah Franklin, Sneja Gunew, Valerie Hartouni, Hilary Hinds, renee hoogland, Joy James, Hannah Landeke, Nina Lykke, Amelia Jones, Kirsten McAllister, Sophie McCall, Maureen McNeil, Adrian Mackenzie, Tom Mitchell, Michal Nahman, Kate O'Riordan, Jackie Orr, Lynne Pearce, Andrew Quick, Celia Roberts, Judith Roof, Anneke Smelik, Lesley Stern, Lucy Suchman, Imogen Tyler, Cindy Weber, Robyn Wiegman, and Liza Yukins. Both Lynn Hershman-Leeson and Frances Leeming have been extremely generous with material relating to their films discussed in chapters 8 and 9. I would also like to express my appreciation for the excellent comments on earlier drafts of this book from Ken Wissoker and from the two readers for Duke University Press. Courtney Berger and Pam Morrison have also been an immense help in the preparation of this manuscript.

I am particularly indebted to Peter Petralia for his calm efficiency and technological know-how in assisting with the final stages of the completion of this book. Without his meticulous work, it would have taken considerably longer to see the light of day. Hilary Hinds and Anna Stacey have lived with this project for many years, and I am extremely grateful to both of them for their enduring patience with its demands on my time and energy. Their good humor continues to be the best antidote to the often unreasonable demands of academic writing.

Introduction

Technologies of Imitation

and the Genetic Imaginary

When we first see the figure of the genetically selected, almost biologically perfect Jerome Eugene Morrow (Jude Law) in *Gattaca* (Andrew Niccol, 1997) he appears from behind a concrete pillar in his wheelchair, in long shot, moving to share the center of the frame with the base of an imposing spiral staircase. The design of the spiral plays upon an association with the now familiar twisted structure of DNA: the double helix. Positioned at the foot of this staircase, flanked by a gray column on either side and rhymed with its green tones and metallic sheen, the figure in the wheelchair is an integral element of a mise-en-scène governed by the pleasures of symmetry and repetition that give a sense of genetics as sequence on the cinema screen. In this vast, bare, stylishly modernist apartment, the wheelchair sits precisely at the place where the spiral staircase begins, so that, at the moment we first encounter Jerome, the visual integrity of the two technologies underscores the cruelty of their paradoxical significance for him. The impossibility of his ascending the very structure that gives a visual presence to the origin of his biological superiority (the double helix) is built into the other technology upon which he depends for his mobility (the wheelchair).

The symbol of the double helix dominates this first encounter between Jerome, the incapacitated Olympic athlete, and his future impersonator, Vincent Freeman (Ethan Hawke), the "invalid" result of a natural conception, with a life expectancy of only 30.2 years. Overlaid with associations of both the discriminatory horrors of past eugenic annihilations and the promise of future genetic cures for disease, the wheelchair and the double helix combine haunting anxieties about the violence of genetic selection with fantasies about engineering improved human beings. The proximity of the figure in the wheelchair to the distinctive twist of the staircase reminds us that, under other historical circumstances, Jerome would have been considered fit only for extermination. Sitting in his wheelchair, surrounded by empty alcohol bottles in his elegant but prisonlike apartment, with its concrete walls and high, narrow windows, Jerome prepares to sell his "valid" genetic code to an ambitious defective "invalid" who seeks to annihilate him in another way—by taking over his identity. As the previously perfect, but now flawed, embodiment of genetic manipulation meets the currently biologically inferior, but soon to be successful, imitator of that lost perfection, the meanings of the wheelchair and the staircase pull in opposite directions: one provides mobility for the body that has lost it, and the other suggests an unbroken flow of energy which seems to animate even static structures through its promise of future genealogical transmissions.

Both the double helix and the spiral staircase are structures of repetition, symmetry, and continuity whose own internal mimetic patterns echo the broader intention of techno-scientific interventions into genetic processes. Just as genetic engineering seeks to imitate biology, so the formal mimicry of the repeating patterns makes each structure self-referential: the doubling within the form reiterates the cloning techniques of genetic science. In *Gattaca*, the design and placement of the staircase perfectly symbolizes the film's exploration of twinning and cloning, as one man attempts a genetic impersonation of another. Like the mirroring helix, Vincent seeks to produce a parallel copy of Jerome's life. While Vincent lives the public life of the successful employee at

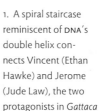

1. A spiral staircase reminiscent of DNA's double helix connects Vincent (Ethan Hawke) and Jerome (Jude Law), the two protagonists in *Gattaca* (Andrew Niccol, 1997).

the Gattaca corporation, Jerome inhabits the lower floor of his apartment, at the bottom of the spiral staircase, secretly preparing transferable genetic samples. The DNA-like structure both connects and divides them: Vincent can move up and down it at his leisure, while Jerome's immobility is underscored by his location at its twisted base (see figure 1). As we wait to see if their joint deception is discovered, the staircase becomes more central to the film's narrative suspense. In a later scene when Jerome is urgently required to answer the door on the upper level (to avoid police detection of the shared crime of identity fraud) he abandons the wheelchair, and the spiral staircase provides the route for his determined ascent. With every new twist and turn of his painful journey as he drags himself up the long spiral, we feel the imprint of this genetic structure on his body and the poignancy of its failed guarantee of physical superiority. Genetically enhanced biology no longer affords him the privileges of his former physical and social mobility.[1]

Not structurally identical to a spiral staircase in a technical sense, but repeatedly associated through visual analogy, the inner ladder of the double helix lends itself to imaginative transformation, as the pairs become stairs in the new spiral architectures linking upper and lower levels as well as past, present, and future generations. The classic visualization of the structure of DNA in popular representations of science displays the twin spirals of the double helix connected by a twisting ladder of adjacent, paired molecules (nucleotides) whose repetitive combinations always include two of the four potential components, adenine (abbreviated as A), cytosine (C), guanine (G), and thymine (T)—whose letters are combined to form the title of GATTACA. But the double helix is not simply a descriptive biological structure that translates neatly into the connective architectures of filmic mise-en-scène to suggest the secrets and deceptions of genetic manipulation. Rather, it is a "figuration," a term Claudia Castañeda uses to refer to something with the "double force" of both "constitutive effect and generative circulation" in ways that speak to "the making of worlds" (Castañeda 2003, 3). If, as Suzanne Anker and Dorothy Nelkin suggest, biogenetics is "a science of the invisible," then the double helix of DNA has given it a visualizable form (Anker and Nelkin 2004, 2). Unlike the double helix, the gene is not an object that can be readily identified in a photograph. The double helix (which typically, though not exclusively, has come to stand in for the gene) has been described as a "powerful and pervasive icon—an almost magical force" (Anker and Nelkin 2004, 1), the fetish concept of our time (Thompson 1995), a "phantom object" (Haraway 1997, 142) and "the only structure linking man to God" (Salvador Dali, quoted in Reilly 2004, xii). As a symbol of a vast domain of semiotic and material practices that might loosely be grouped under the heading of biogenetic sciences, the double helix has now been joined by the images of genetic sequencing, microinjections, and Dolly the sheep that appear regularly in popular representations of techno-scientific interference into biological processes.[2] But none of these, least of all Dolly, quite encapsulates a visual sense of genetic form as the double helix has done. As W. J. T. Mitchell has argued, Dolly's name is more distinctive than her image, which after all, shows a sheep that looks like any other (Mitchell 2005, 15). Nobody could have identified Dolly in a sheep lineup.

Not unlike the Möbius strip, the sign for infinity that presents the endless flow of a repeating pattern, the spiraling continuity of the double helix promises infinite pairings.[3] As the structural enactment of this sense of infinitude, transformed into the fantasy of immortality (Kember 2003, 146), the double helix is a discrete object whose twisted pattern simultaneously presents both the simplicity and the complexity of life. Paul Rabinow argues that the double helix symbolizes a new milieu "beyond the soul," "an imaginary construction with very real effects and potent affects" (Rabinow 1999, 11), and that "the identification of DNA with 'the human person' as a self-evidently synecdochal relationship" constitutes a kind of "spiritual identification" (ibid., 16). The universality of this highly humanistic realm of the spiritual is represented in the rotational motion of this satisfyingly observable display of the miracle of life, which invites the viewer to attempt the impossible task of tracking the winding twists of each helix separately, while watching the two turn in unison. Often animated in rotation, the double helix shows off its own ingenuity of form, suggesting the deeper scientific knowledge about human life hidden in its structures. Referred to as "the blueprint of destiny," the "secret of life," and "a portrait of who we are," the double helix promises to explain the meaning of life and indicates how the biogenetic sciences have been popularly seen as offering "a window on human nature" (Anker and Nelkin 2004, 1).

The iconographic status of the double helix provides the starting point for this book because of its repeated appearance as the sign that makes concrete a techno-scientific field that continues to transform our world invisibly. In *Gattaca*, the spiral staircase evokes a sense of interiority that has come to shape our perceptions of the biological body as geneticized: as a system of cellular processes governed by the activities of genes. Together, the double helix and the spiral staircase make manifest the modern desire to see the body's interior, to externalize its hidden inner forms and make them tangible and imaginable, "to visualize the invisible" (Ostherr 2005, 17).[4] Put simply, the double helix gives architectural shape to the mysterious processes of invisible cellular activity that have come to define the meaning of what Michel Foucault, and many others after him, have called "life itself."[5] This might be thought of as a form of "gene mapping," a particular kind of "spatialization of the body," or a mode of "corporealization" (Haraway 1997, 141). As Donna Haraway suggests, "a gene is not a thing, much less a 'master molecule' or a self-contained code. Instead, the term *gene* signifies a node of durable action where many actors, human and nonhuman, meet" (ibid., 142). Writing of Eduardo Kac's installation *Genesis 1999*, Mitchell sums up the consequent representational problem thus: "the object of mimesis here is really the invisibility of the genetic revolution, its inaccessibility to representation" (Mitchell 2005, 328). In the absence of a recognizable object of photographic reproducibility, the double helix stands in for the gene; it has come to have a ubiquitous presence in contemporary culture as the sign of vitality, immortality, and the future. It spatializes biology and invests that space with the pulse and drive of life.

2. Architecture on
a monumental scale
in *Gattaca* (Andrew
Niccol, 1997).

Although the double helix has become the most ubiquitous sign of the geneticization of the biological body, it is by no means the only genetic motif on the cinema screen. Drawing on an established set of narrative and iconographic conventions of the artificial in science fiction and related genres, the genetically engineered or cloned body is also produced in contemporary film through a number of familiar cinematic tropes of visual deception: twinning, mirroring, doubling, impersonation, and masquerade.[6] Across (and often recombining) the genres of science fiction,[7] the thriller, body horror, comedy,[8] and romantic comedy, narratives of disguise, impersonation, and mistaken identity constitute genetic engineering and cloning as a threat to authenticity, relatedness, individuality, and uniqueness: people are not who they say they are; body surfaces hide deeper truths; you cannot trust what you see.

These threats to traditional markers of the visibility of embodied difference give cinematic life to the artificiality of engineered bodies by spatializing a sense of biological interiority. In a number of films, the giganticism of external architectural structures symbolizes the other side of what Judith Roof identifies as the legacy of the belief that "the increasingly microscopic and infinitesimal . . . [is] . . . the site of fundamental processes" and the place where we find answers (Roof 2007, 33). As Kracauer puts it, "the cinema is comparable to science [in] . . . its preoccupation with the small" (quoted in Landecker 2005, 904). A sense of awe and wonder that matches the body's newly imaginable

minute genetic structures is created by the use of magnificent buildings and spectacular urban locations, such as the corporation's building in *Gattaca* (see figure 2) or the cityscapes of Shanghai in *Code 46*
(Michael Winterbottom, 2003). These contrasting scales of the microscopic and the gigantic produce an aesthetic that displays the ingenuity
of techno-scientific engineering and the seductive formal perfections
of its latest manifestations. Architectural designs create the feeling of
balance, placement, and order within the frame, connecting the newly
discovered genetic landscapes of the inside of the body to the patterns
and structures outside it. Blending architectural forms with an emphasis on sequence, repetition, and symmetry, these films give the interiority of the genetically engineered bodies an integral place within the
mise-en-scène, producing what we might call a geneticized aesthetic: a
distinctly spatialized sense of the gene on the screen.

Insofar as they both seek to imitate life, the cinema and genetic engineering are both technologies of imitation: the first a cultural technology, the second a biological one. This book explores the powerful
combination of their shared mimetic intention. Each displays a fascination with the boundary between life and death, and with the technical
possibilities of animating the human body. If these two technologies of
imitation inaugurate disturbances to our sense of place in the world,
and our connectedness to people and things around us, then what has
been called the "genetic imaginary" spatializes the inner and outer limits of such disturbances.[9] Unlike any other medium or practice, the cinema shares with genetic engineering an imitative but elusive intent that
invites formal interrogation: it brings to life still images and, disguising
its own artifice, invests them with a believable presence on the screen.
As Lisa Cartwright so thoroughly demonstrated in *Screening the Body*
(1995), science and the cinema do not stand in opposite corners of the
ring but rather have co-constitutive technological histories, modes of
perception, patterns of exhibition, and, most importantly here, aestheticized forms of envisioning the human body. This dynamic kinship between science and the cinema has shown not only how the body has
been produced as surface image, but how movement, vitality, interiority, and corporeal activities have been given an imaginative life. Importantly, Hannah Landecker has argued that since its inception, the
cinema and science have been mutually informing practices. Her own

concern with early cellular films and the use of microcinematography in the production of particular conceptions of biological life demonstrates perfectly that the entwining of the visualizations of "life as such" in biogenetic science and the cinema is the latest set of intersections in a long history of convergent technological practices.[10] *The Cinematic Life of the Gene* plays on the converging desires to imitate life in science and in the cinema, placing this study firmly within feminist cultural studies of new forms of visualizing life in both spheres. It explores how the language of film animates a sense of the geneticized body, how it gives visual form to the invisible gene, and how these fictions are structured by anxieties both about the changing meaning of cinematic form and about the transformation of the nature of the human body and its sexual and reproductive future.

The chapters which follow offer close readings of a series of theoretical and film texts whose selection was governed by a unifying set of interdisciplinary conceptual concerns with the ways in which a social imaginary combines psychic dynamics with cultural formations. My starting point for thinking about the cinematic life of the gene is to interrogate how the genetic imaginary constitutes a set of very tangible anxieties surrounding the reconfiguration of the boundaries of the human body, the transferability of its informational components, and the imitative potentialities of geneticized modes of embodiment. In this investigation, I define the genetic imaginary as the mise-en-scène of these anxieties, a fantasy landscape inhabited by artificial bodies that disturb the conventional teleologies of gender, reproduction, racialization, and heterosexual kinship. In the genetic imaginary, we see the invention of posthuman life forms whose histories can be manipulated and whose futures might be extended, but who threaten to exceed the controlling gaze of scientific technologies and thus continuously trouble their authority. Concerns about the destabilization of traditional markers of difference and privilege combine with those about the introduction of the unnatural and the inauthentic at both the biological and representational levels: genetically engineered and cloned forms of life, and those materialized by the artifice of digital cultures. While anxieties about the loss of authenticity and nature are hardly new (indeed, they permeate modernity and postmodernity), and fears about the separation of sexuality from reproduction or about the visibility of racial differences have not been inaugurated by forms of genetic manipulation alone, they are amplified and reconfigured in the genetic imaginary through a particular preoccupation with imitation, disguise, and copying in which new modes of facilitating the technological legibility of identity blend with new codes of deception in the age of replication (see Smelik and Lykke 2008).

The imaginary is an ambiguous term, and it is worth pausing briefly on what it has come to mean in some strands of feminist cultural studies, and to define more specifically how I shall be using it in this book. The genetic imaginary brings center frame the question of fantasy in debates about technical innovation.[11] With the notable exceptions of Roof's (1996b) analysis of how changing notions of biological and cultural reproduction destabilize the foundations of the symbolic order, Sarah Kember's (1998) exploration of the hopes and fears surrounding reproductive and information technologies, and Haraway's work on gene fetishism,[12] there is still very little work linking questions of desire, fantasy, and subjectivity to the study of the reconstitution of the body through genetic engineering, and there is almost none which brings psychoanalysis to bear on debates about how science produces particular cultural desires in this context (Haraway 1997, 131–72).[13] What Haraway calls "gene fetishism" involves "mistaking *heterogeneous* relationality for a fixed, seemingly objective thing" (ibid., 142). According to this argument, "the gene as fetish is a phantom object, like and unlike a commodity"; it involves a kind of "'forgetting' that bodies are nodes in webs of interrogations, forgetting the tropic quality of all knowledge" (ibid.). Combining Marxist and Freudian notions of fetishistic denial and disavowal, Haraway suggests that "gene fetishism involves . . . elaborate surrogacy, swerving, and substitution, when the gene as the guarantor of life itself is supposed to signify an autotelic thing in itself, the code of codes" (ibid., 147). Haraway describes her commitment to "excavate" something like the "technoscientific unconscious, the processes of formation of the technoscientific subject, and the reproduction of the subject's structures of pleasure and anxiety" (ibid., 149). This combination lies at the heart of my use of the notion of a genetic imaginary whose cultural formations draw so powerfully on psychic investments. The dynamic flow of associations across these interconnected fields is tracked throughout this book.

It is this move that the collage animation film, *Genetic Admiration* (Frances Leeming, 2005), discussed in detail in chapter 9, makes so brilliantly. Written, directed, and produced by Leeming, a Canadian independent filmmaker, this short film uses the free-associative methods of surrealist/Dadaist collage to expose the unconscious motivations of scientific projects, and connects them to the iconographies and narratives of popular culture. The film questions the desires behind the spectacular science of genetic engineering and cloning displayed to the public through mass-entertainment genres in which a woman's body becomes the vessel for masculine adventure narratives of a scientific and cinematic nature. The cut-and-paste techniques on the screen deny us the pleasures of the smooth transitions and fluid mutations found in both hidden filmic suturing and invisible genetic practices. Disturbing the immaculate surface of familiar visions, the film interrupts the flow of logic between ways of seeing and ways of knowing. In chapter 9, I read this film as an enactment of the genetic imaginary.

This book offers a study of the genetic imaginary through the consideration of a series of cultural disturbances: to the biological foundations of embodied difference; to the visual intelligibility of the human; and to the continuity of singularity, individuality, and authenticity. The prospect of techno-scientific interference in the genetic processes of human life generates these disturbances in multiple ways: it threatens to break in upon nature and interrupt its imagined autonomy; it repositions kinship and reproductive arrangements within a technological frame; it throws traditional couplings into unnatural and alarming disorder; and it destroys the composure of settled boundaries and distinctions, however fictional their original basis may be. In the chapters that follow, I consider how we might conceptualize the elusive nature of geneticization, exploring the claim that the genetic imaginary is distinguished by this discrepancy between a powerful affective force and an intangible presence, and offering an account of the paradoxical absent presence and visible invisibility of the gene in contemporary culture.

Throughout the book, the genetic imaginary refers to the organization of cultural fantasies in ways that are part of the psychic production of subjects. Although in cultural studies generally, the imaginary is sometimes used loosely and interchangeably with the concept of the cultural imagination and has become semidetached from its psychoanalytic origins, for this project the term still carries with it important traces of psychoanalytic thinking which remain central to my analysis.[14] Unlike the imagination, with its roots in philosophical and aesthetic conceptual traditions, the imaginary implies a set of structures for the production of subjectivities with the power to draw upon and reproduce unconscious attachments. While the imagination refers to patterns of emotional and artistic connectivity, the imaginary refers to the fears and desires organizing a particular repertoire of fantasies that have a deeper, often indirect, set of cultural investments and associations.

The genetic imaginary is explored here in terms of three structuring desires: to imitate life in both science and the cinema; to secure identity as legible through screening technologies; and to anchor embodied difference by making it stable, predictable, and visible. Across these desires runs a set of anxieties integral to, but not exclusively generated by, genetic engineering and cloning: the detraditionalization of heterosexual reproduction and the queering of biological processes; the problem of identity theft and genetic impersonation; and the informationalization of both the image and the body in contemporary culture. Leo Bersani and Ulysse Dutoit (2004, 1) write that "a major virtue of the visual arts is their capacity to make the invisible visible." It is precisely the desire to realize this promise that drives the genetic imaginary, particularly its cinematic forms. The films discussed in this book are all preoccupied with this paradox: the impossibility of seeing someone's genes despite the ubiquitous presence of genetic discourse. Each of the chapters that follow this introduction elaborates a particular conceptual dimension of this paradox that drives the genetic imaginary.

The book offers a series of readings of theories and films (and of each through the other) that consider the ways in which cinematic and genetic techniques share a particular imaginative charge at the present moment. The cultural reconfiguration of life with which this book is concerned occurs with particular intensity in what I rather loosely refer to as "the decade of the clone": the mid-1990s to the mid-2000s is a decade of proliferation of debates about and representations of cloning and genetic engineering. Though the precise boundaries around the decade are debatable, there is something distinctive about the extent to which it witnessed multiple announcements of groundbreaking discoveries and experiments in the field of biogenetics. A number of breakthroughs and threshold-crossing innovations were heavily publicized during this period, proclaiming the dawning of a new era of techno-scientific possibilities through genetic manipulation: most notably the cloning of Dolly the sheep in 1996, the establishment of lines of embryonic

stem cells in 1998, the completion of the mapping of the human genome in 2000, and the first successful cloning of a human embryo in
2004 (a claim that was later retracted).[15] As a result of this intensification of public debate and media visibility, genetic discourse now pervades everyday culture and has become part of all kinds of medical,
legal, commercial, familial, and virtual encounters. In accounts of the
discourse's emergence, the decade of the clone marks the time during
which the word "genomics" entered public and media debates with
an unprecedented force, through the advent of public engagement with
cloning and related techniques.[16] While the word "genetics" refers to
the study of genes, genomics means the study of an organism's entire
genome, including genetic mapping and the DNA sequence of the whole
entity; it thus involves the combination of genetic techniques with sequencing through information and computing technologies.[17] This more
expansive and ambitious move is not just a technical shift but symbolizes a whole set of discursive and performative changes in the way
techno-science could claim to understand life. As José van Dijck argues in *Imagenation* (1998), the Human Genome Project is much more
than a concerted effort to produce an inventory of the human genome
as information: it has entailed "the development, distribution and implementation of a way of thinking about human life" (ibid., 119); I shall
discuss this further in chapter 3.

The flood in the past twenty years of films about genetic engineering
and cloning intensified in the decade of the clone. One study suggests
there were five times as many cloning films released in this decade compared with the previous one;[18] Mitchell (2005) suggests that there have
been well over a hundred such films released in the last two decades
alone.[19] But this book is not intended as a comprehensive, exhaustive,
or even necessarily representative account of these films. It is not a sociological survey, nor is it a history of a particular genre or subgenre.
The focus on the cinema in this study is motivated by conceptual rather
than empirical concerns; it is not meant to imply that other media are
insignificant in the production and circulation of genomic discourses.
My interest here is not to capture the meaning of these discourses from
media content, regardless of formal elements; it is, instead, to explore
the connections between the ephemerality of cinematic and genetically
engineered forms of life, and to interrogate the desire for visual certainty generated by such insubstantiality. In other words, it is precisely

the imaginary vitality of their convergent formal properties that interests me here.

The book divides into three parts with the following overarching concerns: the first part addresses the genetic reconfiguration of sexuality,
reproduction, and kinship;[20] the second explores genetic forms of mimicry and impersonation; and the third compares the changing status
of the body and the image in terms of their reproducibility. The preface sets out the rationale for the structure of the book; however, the
theoretical chapters and the film readings not only speak to each in
the sequence in which they appear, but they are also in dialogue through
a more spiraling logic that cuts across them.

The fears about the queering of the social through a technologized
reproductive biology rehearsed by Jean Baudrillard, discussed in chapter 1, are most obviously put into dialogue with the reading of *Alien:
Resurrection* in chapter 2. But the queer associations of scientific interference in human biology are also present in other films discussed in
the book: *Gattaca* plays with genetic impersonations of the clandestine
couple through a classic homosocial triangulation with a female third
term who may seem like more than one sort of cover (chapter 5); *Code
46* embeds its flirtation with the queer connotations of warped kinship within its stylized fleeting moments of sophisticated metrosexual
encounter (chapter 6); and *Teknolust* undoes some of those traditional
couplings and laughs at the nostalgic longing for an unblended gender
or an unmediated biology, even as it seems to reinforce the assignment
of homoeroticism to the place of synthetic reproduction (chapter 8). A
number of the film readings also present the prospect of a cinema without Oedipus in ways that address Baudrillard's (2002) fears about unnatural kinship futures.[21] The future of the cinema (as well as that of
the subject) is haunted by the prospect of post-Oedipal reproduction,
from the immortal Ripley's transgenic and transgenerational kinship
(she "mothers herself") through the technologization of the incest prohibition in *Code 46* to the cinematic substitutions for the production of
desire in the absence of Oedipus in *Teknolust*. Finally, *Genetic Admiration* stages absurd post-Oedipal kinship scenarios, which blur the distinctions between scientific discovery and science fiction (chapter 9).

Similarly, the film readings of the threat and desirability of artifice
in *Gattaca* and *Code 46* relate most directly to theories of mimicry and

masquerade, but other film chapters concern mistaken, stolen, or fabricated identities. In both *Species* (chapter 3) and *Teknolust*, the heroine's sexual irresistibility is a programmable and performable quality, and genetically engineered biological instinct and digitally replayed sexual seductions push the idea of femininity as artifice to new limits. The centrality of the woman's body to the production of both scientific knowledge and cinematic pleasure is typified in *Species*, enacted in *Teknolust*, and deconstructed in *Genetic Admiration*. The desire for perfection and the fear of impurity, hybridity, freaks, and monsters are also highly racialized scenarios of mimesis: the whiteness of the feminine embodiments of genetic artifice (Ripley in *Alien: Resurrection*, Sil in *Species*, or the clone triplets in *Teknolust*); the problem of inappropriate genetic mixtures in the globalized multiculturalism of *Code 46*; and the mockery of the white adventure narratives of colonial expansion and conquest informing both the pioneer spirit and scientific endeavors sutured into the rough edges of the collage animation of *Genetic Admiration*.

The place of the genetically engineered white body in these films highlights a paradox. White bodies have conventionally been a sign of desirability, both in terms of symbolizing racialized purity and of displaying ideals of physical perfection in the history of the cinema;[22] but if whiteness is achieved through some kind of interference at the level of the genetic code, then both purity and perfection are surely compromised. Cloning films typically concern the creation of white ideals, but these figures are often presented as ruined or spoiled in some way that undermines such idealization (for instance, Jerome's fading traces of sculpted desirability in *Gattaca*, or Ripley's connection to monstrous births on display in *Alien: Resurrection*). The lack of humanness in the white protagonist (the hyperanimated physical strength and beauty in *Species*, or the painted geisha look of the automated sexual performer in *Teknolust*) constitutes the clone figure as a "supplement" that reflects back on the limits of the original subject in question (Battaglia 2001, 496). These white clones embody both a hyperbiological human—what Eugene Thacker (2004) calls a "rebiologised nature"—and its artificial other. Since many of these cross-species enhancements involve dangerous outcomes (Ripley's violence, Sil's strength, and Ruby's contagious virus) the problem of the white body as the sign of the human seems to need endless rehearsal in these films. Both too artificial (having lost touch with nature) and not pure enough (threatened by unnatural mixtures), the white body is the site of genetic amplification that ultimately may serve only to expose its lack of originality and authenticity, or to expose it to the potential violent consequences of mixed biologies.

The discussion of whether cloning inaugurates the loss of the body's aura in the Benjaminian sense (what we might call the double take, or shock, generated by mechanical and genetically engineered reproduction) is read through the desire for authenticity as individualized affect not only in *Teknolust* and *Genetic Admiration* but also in *Alien: Resurrection*, *Gattaca*, and *Code 46*. Questions of what constitutes the human in these so-called posthuman times (Hayles 1999) are explored through the ways in which these films attempt to locate authenticity in either sexuality or the power of affect: Ripley's compassion in *Alien: Resurrection*; the combined willpower of the two Jeromes in *Gattaca*; the strength of heterosexual desire in *Code 46*; and romantic love and artistic expression in *Teknolust*.

To some extent, each film reading is in dialogue with all three theoretical speculations (in chapters 1, 4, and 7): the process of mourning the demise of the human in clone cultures; the imperceptible distinction between copies and originals in practices of impersonation and mimicry; and the idea of a lost bio-aura. All the chapters attempt to conceptualize that place of authenticity beyond mechanization, the disciplinary regimes of modernity, and the reproducibility of commodity culture. This fantasy of escaping technology either psychologically, geographically, or sexually runs across all the films discussed, but with the advent of biotechnology (where biology becomes technology), such a place becomes increasingly difficult to imagine.

The Cinematic Life of the Gene places feminist debates about the current reproducibility of biological matter through genetic engineering in dialogue with those about the reproducibility of the image through photographic and cinematic techniques, and with corresponding debates about the recent shift from analog to digital techniques. In addressing itself to the new "modes and codes" (Smelik and Lykke 2008, xiii) of contemporary technologized forms of life, this book combines psychoanalytic concerns about the changing psychic demands of technoscientific inventions with cultural analysis of the reconstitution (or indeed erosion) of the body and its materiality. Anxieties about the endurance of human individuality and authenticity have a long history, but the problem of the visibility of difference has taken on a particular force in both science and the cinema since the mid-1990s. Troubling our desire to read identity from surface appearance, looking the same and being the same have become separable. If the very technology that seemed to guarantee our individuality at the genetic level also promised the extraordinary possibility of our self-duplication, then the cultural imaginary might well struggle to defend itself against the most extreme implications of such a paradox. If genetic makeup is extractable, for example, how can it be an expression of singularity? If identity can be translated into information, could this not be circulated, reproduced, even borrowed or imitated in ways that trouble the fantasies of unique individuality—which, as Katherine Hayles (1999) and many others have argued, lie at the heart of the liberal humanism of Western cultures? What is geneticized embodiment, and what new forms of literacy are required to render such a body legible?

This book addresses these questions by moving between theoretical and filmic texts to consider how the animation of cellular life at the genetic level is produced in the cinema at a moment where the mutability of the body coincides with the mutability of the image, in both cases threatening particular diachronic continuities. It is precisely the tension between the ambiguity of the meaning of the gene and the desire for certainty of form and vision that unites the very different theories and films discussed in this book. In grappling with the cultural disturbances presented by such ambiguities, the following chapters read across the particular repetitions that constitute the forms of the genetic imaginary, searching for ways to loosen their tight grip on our psyches and our visions.

[1]

Sameness Ad Infinitum

Human beings have always done this [changed the rules at the risk of coming to grief] in the symbolic order, now they will do it in the biological order. . . . [T]he incredible violence of genetic simulation [cloning] . . . is the final phase of a process our modern technologies have simply speeded up: the process of an ideal counterfeiting of the world, allied to the phantasm of an immortal recurrence. In other words, the perfect crime: the work of finishing off the world, for which we now have to undergo a process of mourning. BAUDRILLARD, *Screened Out*

[1]   The Hell of the Same

Cloning, Baudrillard, and the Queering of Biology

Cloning presents us with the potential to undo what we have come to understand as the uniqueness of human life, the finality of death, and the place of sexuality and reproduction in the formation of culture. Despite the obvious overlap with psychoanalytic concerns (such as twinning, doubling, alterity, the death drive, the end of Oedipus, the uncanny, mourning and melancholia, intersubjectivity, womb envy and abjection) the subject has attracted surprisingly little commentary from such theorists.[1] We might have expected recent developments in genetic engineering to have generated a plethora of debates about the unconscious desires underpinning these new scientific techniques and governing our responses to their increasingly evident social presence.[2] But in the burgeoning literatures across the humanities and social sciences there are strikingly few accounts of the psychic or affective dimensions of cloning and genetic engineering, nor indeed of how the Oedipal foundations of culture are currently being threatened by the existence of these potential forms of artificial life.

In contrast, Jean Baudrillard is one of the few to theorize the unconscious drives behind cloning as a scientific project. Specifying the psychic violence it poses to the structural underpinnings of the symbolic order, Baudrillard asks: What are the phantasms "underlying the whole genetic project?" (2002, 198). Sitting somewhere between cultural theory and science fiction, his deliberately provocative and apocalyptic answers caution against the ways in which these postmodern imitations of biological life are destroying the foundations of both the human psyche and the symbolic order through the current social realignments of the human and the technological, and the reconfiguration of life, death, and sexuality.[3] This focus makes Baudrillard's writing on cloning an obvious starting point for a book concerned with elaborating the dynamics of the genetic imaginary.[4]

This chapter takes Baudrillard's writing on cloning as a symptomatic focus for mapping the conceptual mise-en-scène of this imaginary in which the psychic forces of techno-scientific innovation might be read through particular cultural forms. Part of a wider critique of the cultural economies of postmodernity, Baudrillard is concerned here with the ways in which genetic engineering threatens unconscious processes of subject formation (or, as he puts it, perhaps requires us to dispense with both the notion of the unconscious and of the subject altogether). Crystallizing the profundity of the current cultural interference generated by genetic engineering, he asks: "Is it possible to speak of the soul, or the conscience, or even the unconscious from the point of view of the automatons, the chimeras, and the clones that will supersede the human race?" (Baudrillard 2000, 23).

But to read Baudrillard's theories as indicative of the shape of the genetic imaginary more generally may also provoke scepticism from some readers unconvinced by the universalizing and often inflammatory rhetoric for which he has become famous (see Edelman 2005). Many of his critics would dismiss his interventions as mere paranoia and polemic.[5] For my purposes here, though, this is precisely what makes his writing so significant, since his hyperbole is performative of many of the underlying psychic terrors driving such an imaginary. In mapping the deepest cultural anxieties about cloning as part of an analysis of the "phantasms underlying the whole genetic project" (ibid.), Baudrillard's work reiterates the associative moves which constitute the genetic imaginary. Building on the force of Baudrillard's discursive excess, my readings of his work trace the "transmission of affect" (Brennan 2004) generated by the advent of cloning. Baudrillard's work is perfectly suited to an indicative reading of the genetic imaginary because of the ways in which it rehearses the affective disturbances of cloning and genetic engineering.

This rest of this chapter explores the landscapes of such an imaginary through a series of close readings of the discursive production of *sameness* in the account of "the genetic project" in Baudrillard's work. Drawing on my previous work on how abjected desires and bodies become associated with a deathly proliferation of sameness (see Stacey 1997), I examine Baudrillard's claims that, in dispensing with sexual difference as the embodied guarantee of the humanness of reproduction, cloning threatens to reverse evolution and announce the end of the species as we know it. The fear of sameness, much debated in queer and feminist theory, is pivotal here to the desire to re-privilege heterosexual difference, even as genetic discourse simultaneously enables its most profound denaturalizations. As I discuss below, for Lee Edelman, Baudrillard's "paradigm of sameness" implicates his work irredeemably within the heteronormative logic of what Edelman calls "reproductive futurism" (2005, 61). My reading seeks to make explicit how the genetic imaginary casts the current problem of sexual difference as a simultaneously technical, psychic, and social one, implicating the new possibilities of genetic interference within the transformation of social structures and the loss of the legitimizing power of foundational sexual categories.

Baudrillard's writing on cloning appears in *Screened Out*, published in 2002; perhaps appropriately, however, there are numerous previous incarnations of these ideas. His most elaborated challenge to cloning and other forms of genetic engineering is his essay titled "The Final Solution," published in 2000 in *The Vital Illusion*.[6] With its direct reference to the Nazi plan to exterminate the Jews of Europe—an association which continues to haunt contemporary genetics, despite attempts to recast the latter's ethical and political credentials and genealogy—this essay pushes the logic behind the cloning project to its discursive limits in an attempt to reveal what Baudrillard considers the self-destructive absurdities of its impetus. In this short piece, Baudrillard rehearses the violent symbolic implications of the new genetic imitations of life and elaborates what he sees as the phantasms driving such a project.

In his claim that as a fantasy of immortality, cloning is ultimately a "fatal strategy" powered by the death drive, Baudrillard places center stage the threat posed by genetic engineering to traditional configurations of sexual difference and embodied heterosexual reproduction (Baudrillard 2002, 197–99). For Edelman, Baudrillard's arguments about cloning here reveal him to be an advocate of "reproductive futurism"—the framing of the political field as defined by terms "that impose an ideological limit on political discourse as such, preserving in the process the absolute privilege of heteronormativity by rendering unthinkable . . . the possibility of a queer resistance to this organising principle of communal relations" (Edelman 2005, 2). Edelman rightly exposes a logic generated in Baudrillard's polemic through which the subject not only has much to fear, but might already be lost, since the foundational function of sexual difference is to be replaced both biologically and culturally by a proliferation of sameness, imitation, and artifice. In his declaration that the age of the clone, or what he calls "the degree xerox of the species," means we are approaching the end of life as we know it, Baudrillard reads genetics as simply the latest manifestation of the desire to reinvent the "rules of the game, at the risk of coming to grief" (Baudrillard 2002, 198).

Placing cloning within the more general framework of his polemic against the simulacra and simulations of postmodern culture, Baudrillard sees genetic engineering as the biological equivalent of the cultural imitations that surround us in contemporary Western society: "isn't virtual reality as a whole an immense cloning of the so-called real world?" (ibid., 200). Cloning thus blends the biological and cultural through the structures of communication that can now only be mobilized for repetition. As an *écriture automatique* of the ready-made, cloning is conceived here as writing from beyond the grave through the live body, offering vicarious agency to the dead. As such, it threatens to replace both the authenticity of the body and the spontaneity of creativity with the ghostly replication of multiple automatons, whose illusory individuality becomes the exchange value of the markets of new technologies.

With the advent of cloning, according to Baudrillard, it is not simply that the economic relations of capitalism require social reproduction through culture, but that biological and cultural reproduction have now become unified and serve the same futile end. If it seems that culture alone will preserve us from "the hell of the same" (1993, 113) of cloning, Baudrillard cautions us that "in fact, the reverse is true. It is culture that clones us, and mental cloning anticipates any biological cloning" (2000, 25). In this sense, the logic of genetic engineering is already present in the "social register, where what the system produces and reproduces are virtually matching beings, beings substitutable for each other, already mentally cloned" (2002, 199). Since cloning is merely the logical biological extension of the "culture of the copy" (Schwartz 1996), we have already been programmed to accept cycles of replication. Baudrillard argues that "social cloning, the industrial reproduction of things and people, . . . makes possible the biological conception of the genome and of genetic cloning, which only further sanctions the cloning of human conduct and human cognition" (Baudrillard 2000, 25).

More than this, he suggests that cloning is an "infinitely more subtle and artificial prosthesis than any mechanical one" (Baudrillard 2002, 200). The desire for cloning is an extension of previous accumulated imaginings: "everyone may at some time have dreamt of duplicating or multiplying himself to perfection, but this is, precisely, a fantasy: it is destroyed when you attempt to force reality to conform to it. Everything—utopias, transparency, perfection—becomes terrifying as soon as it is turned into reality" (ibid., 199). In this sense, if cloning builds on existing phantasms which have haunted Western culture (so keenly wedded to singularity and authenticity as the anchors of subjectivity), it is the literal manifestation of what was up until now a "merely mental or metaphysical" fascination: "when this fated character of the Same . . . becomes materially inscribed in our cells . . . science itself becomes a *'fatal strategy'"* (ibid., 198–99, emphasis added). As the ultimate consequence of all modern technologies, cloning both repeats the accumulated errors of misguided history and leads to unprecedented violence that threatens the very foundations of the symbolic order. Baudrillard's apocalyptic rhetoric of "fate" and "fatality" gives force to this discourse of historical inevitability, anchored in the marriage of psyche and biology.

Cloning thus conceived is the result of a nostalgic longing for a "return to . . . the Same" and can be understood only as the triumph of the death drive: "this dizzying temptation to return to annihilation in the eternal repetition of the Same, to go back beyond the biological revolution of sex, back beyond death" (ibid., 197). This declaratory rhetoric gives further force to the sense of the inevitability of cloning, however undesirable. The breadth of this historical sweep (encompassing past and future so readily) intensifies this pessimism as it condemns the similarity in forms of biological and cultural reproduction. For Baudrillard, it is the death drive which propels such insanity forward into inevitable oblivion. Only that drive, he says,

> would impel sexual beings towards a presexual form of reproduction (in the depths of our imagination, moreover, is it not precisely this scissiparous form of reproduction and proliferation based solely on contiguity that for us *is* death and the death drive?). And what, if not a death drive, would further impel us at the same time, on the metaphysical plane, to deny all otherness, to shun any alteration in the Same, and to seek nothing beyond the perpetuation of an identity, nothing but the transparency of a genetic inscription no longer subject even to the vicissitudes of procreation? (Baudrillard 1993, 114)

If science has become a "fatal strategy" through which the desire for death masquerades as the desire for life, then fantasies of immortality and the death drive collaborate in the cloning project. Contrary to the common-sense assumption that the desire for death and the desire for immortality might pull in opposite directions, Baudrillard suggests that they are united in the figure of the clone, since the form of immortality longed for is itself a kind of species suicide: the drive for immortality and the death drive "are in play simultaneously, and it is possible that one is nothing but a variant of the other, nothing but its detour" (Baudrillard 2000, 27). Rather than enacting a longing for life, cloning represents the opposite: a longing for a return to nondifferentiation, sameness, and self-annihilation. He formulates this in another way, saying that cloning represents "the forgetting of death"—the possibility that we shall not die (ibid., 5). Perhaps this is the greater terror: "at the slightest hesitation in the fight for death—a fight for division, for sex, for alterity, and so for death—living beings become once again indivisible, identical to one another—and immortal" (ibid., 5–6). In the genetic imaginary, the clone figure thus embodies both the death drive which impels a culture toward such a genetic science, and the desire for immortality which such techniques, like others before them, promise to deliver.

The fantasies of immortality motivating the project of cloning, Baudrillard suggests, return us to an unhealthy longing for biological immortality from which we had previously escaped: "blindly we dream of overcoming death through immortality, when all the time immortality is the most horrific of possible fates" (2000, 6). The force of his rhetorical case against cloning in effect rests upon the invocation of a familiar (and highly conventional) version of evolutionary biology: "The evolution of the biosphere is what drives immortal beings to become mortal ones . . . Encoded in the earliest life of our cells, this fate is now reappearing on our horizons . . . with the advent of cloning." The threat of cloning is located within an evolutionary teleology of biological and psychic progress from undifferentiated life to advanced human sexual reproduction: "Contrary to everything that seems obvious and 'natural,' nature's first creatures were immortal. It was only by obtaining the power to die, by dint of constant struggle, that we became the living beings we are today" (2000, 6). A culture that endorses cloning is cast as a diseased return to the "primitive" moment of futile biological cycles of cell division and replication; cloning reverses the previously natural order of evolution, whose drive toward mortality saved us from the horrors of immortal self-replication:

> By way, paradoxically, of science and progress, we are now quite simply eradicating the greatest revolution in the history of living beings, the transition from protozoan, bacterial, undifferentiated cell division—the immortality of the single-cell organisms—to sexual reproduction and the inalienable death of every individual being, and replacing this with the biological monotony of the earlier state of affairs, the perpetuation of a minimal, undifferentiated life, for which we have perhaps never stopped yearning. (Baudrillard 2002, 196–97)

Biology here is the realm of the problematic genetic possibilities of which Baudrillard is so critical, as well as the place of rhetorical counterpoint from which to judge these new technological crimes against nature. The evolutionary push toward differentiated cells is seen by Baudrillard as synonymous with the emergence of heterosexual reproduction, blended in the bodies of the human male and female—whose obvious biological differences are the anchor of a healthy society. In this narrative of evolutionary progress, the repetition of dichotomized sex difference through heterosexual reproduction places the foundation of culture firmly within biology. Indeed, our psychic sanity relies precisely upon our mortality, which can be guaranteed only by the natural signifying power of heterosexuality as the incontestable and visible marker of the human. In placing sex difference in such a foundational position, Baudrillard makes it the exclusive source of difference necessary for establishing and sustaining the singularity and authenticity of human life and its symbolic formation.

For Baudrillard, the cloning ideal involves going against the healthy evolutionary impetus by which human beings have striven to "wrest the same away from the same—to tear themselves free from this kind of primitive entropy and incest" (ibid., 197).[7] Cloning represents only the "cancelling of differences" and the "manufacturing" of "entropy with information"; it is the final solution insofar as it represents the revenge on "mortal and sexed beings by immortal and undifferentiated life forms" (Baudrillard 2000, 8). The whole genetic project could be seen to be motivated by the desire to reactivate an unhealthy relation to sameness, defined here as identical biological cell formations with an endless life cycle of multiplication rather than death. "Involution" replaces "evolution," as we consider the possibility that "this final solution toward which we unconsciously work is not the secret destination of nature, as well as of all our efforts" (ibid., 9). As Baudrillard sums up: "This is our clone-ideal today: a subject purged of the other, deprived of its divided character and doomed to self-metastasis, to pure repetition. . . . No longer the hell of other people, but the hell of the Same" (Baudrillard 1993, 122).

Cloning threatens to return us to our "primitive" biological state of nondifferentiation, according to Baudrillard. With all its colonial connotations, this psychoanalytic notion that the biological sameness of cloning returns us to a regressive desire for immortality motivated by the death drive hinges on the motivating presence of an evolutionary unconscious that shapes knowledge production. We might thus trace the loss of a progressive drive toward difference through an evolutionary logic embedded within the human psyche, and find instead the triumph of a nostalgic longing for sameness. With the unconscious firmly located within the evolving (or involving) history of the species, the psychic and the biological confirm each other's need for difference. Biology, it seems, has steered history toward a healthy relation to difference, whereby undifferentiated immortal life forms became sexed mortal beings; but, like the human psyche, it has been haunted by a self-destructive desire for annihilation, entropy, and merger.

Biological difference (here read as sex; see Butler 1990) is mapped onto psychic difference (survival), as sameness in each is cast out as the undesirable other of life-affirming mortality. Sexual difference here becomes the result of evolutionary progress. Baudrillard suggests that "the evolution of the biosphere is what drives immortal beings to become mortal ones." The emergence of sexed cells is tied to the necessary drive toward mortality: "They move, little by little, from the absolute continuity found in the subdivision of the same—in bacteria—toward the possibility of birth and death. Next the egg becomes fertilized by the sperm and specialized sex cells make their appearance." In this evolutionary model of progress toward biological sexual difference, he argues that "there is a shift from pure and simple reproduction to procreation: the first two will die for the first time; and the third will be born. We reach the stage of beings that are sexed, differentiated, and mortal" (Baudrillard 2000, 7).

According to this reading, genetic engineering threatens to return us to the biological malformation of regressive sameness, of which, this logic suggests, incestuous desire is only the most obvious form. Placing the desire for cloning within an insane landscape of endless cycles of biological malformation, Baudrillard suggests that "we're going to be able to clone ever more sheep and so make more and more animal feed with which to feed ever more mad cows" (Baudrillard 2002, 196). Connecting the "Crazy Cow" and "Baby Dolly" through an image of cloning as diseased, he contends that "cloning is itself a form of epidemic, of contagion, of metastasis of the species—of a species in the clutches of identical reproduction and infinite proliferation, beyond sex and death" (ibid., 196).[8] Baudrillard sees cancer as the quintessential analogy for cloning since cloning, like a malignancy, involves cells which have forgotten to die, and "forget *how* to die"; if the cell becomes cancerous, it "goes on to clone itself again and again" (Baudrillard 2000, 5). Cloning promises a world in which we might all join the deviant cells of cancer that proliferate and threaten our mortality by forgetting to die. The "inverse peril" of the fight against death is "the fight for death" (ibid., 5–6).

The anxieties which previously turned cancer into the most dreaded of diseases (see Sontag 1977) through an associated imagery of the invisible other within (Stacey 1997)—that metastasizing body which has hidden its presence from its host—are invoked here through the haunting figure of the clone (the other without), who is placed within a similarly ghostly realm through its artificial imitation of cell division. The notion of the "metastasis of the species" magnifies the malignancy of cancerous growth to the scale of the universal: the whole of humankind is turning against nature, as undifferentiated cell proliferation extends into undifferentiated imitation of the human race. Species (the word comes from the Latin *specere*, meaning to look at or behold) should be observably distinct from one another, but with cloning the original and the copy become indistinguishable. Just as cloning articulates a desire to move beyond sex and beyond death, so it is a return to the primitive, the incestuous, and the same. Such "involution" dispenses with the necessary foundations of the human race, which place sexual difference and embodied reproduction at the heart of life itself. The evolutionary progress away from sameness and toward difference, away from self-replication and toward biological reproduction, is anchored in the imbrication of sexuality and reproduction which defines the distinctiveness of the human. According to this trajectory, phylogenesis becomes tautogenesis—in other words, plant reproduction (based on cell division into two more of the same) becomes human reproduction (the birth of the same). Governed by tautology (the logic of the same), tautogenesis supplants the evolution of the species with a new law of regressive division masquerading as the real thing: heterosexual reproduction.

In aligning cloning with disease, Baudrillard condemns the artificiality of genetic imitation through the haunting trope of cancerous proliferation: malignant, abnormal cell growth. Just as cancer is a disease in which the body fails to recognize the otherness of undifferentiated cells, so cloning is a metastasis insofar as it offers an undifferentiated copy which mimics an original. A disease of the self in which the cells turn against their host body and take over its organs by passing as normal cells and hiding their deviant nature, cancer as metastasis masks its own imitative intentions. Since cloning involves artificially stimulated cell division, and cancer is the proliferation of cells against the body's better nature, the biological analogy takes force through the fear of unnatural processes distorting an otherwise natural system of molecular regeneration. As I have argued elsewhere, the designation of cancer (the C word) as the disease of the secret hidden within the self shares the affective force of the stigmatized category of sexual deviance endlessly spoken through euphemism (the L word): each is degraded through its associations with a failure to recognize (and desire or destroy, as appropriate) difference and with the consequent deathly proliferation of sameness (Stacey 1997). Here, cancer is the pathological equivalent of reproduction without sexual difference, which for Baudrillard represents an endless repetition of the same.

For Baudrillard, cloning annihilates the centrality of sexual difference to embodied reproduction and thus produces the threat of unregulated excessive sameness: "the key event here is a liquidation of sexual reproduction and, as a result, of any differentiation of—and singular destiny for—the living being (Baudrillard 2002, 196). Once again criticizing genetics, yet drawing on its discursive power to naturalize heterosexual reproduction, Baudrillard uses biology to privilege sexual difference as the necessary parental division which guarantees the future of the human. He suggests:

> Cloning . . . radically eliminates not only the mother but also the father, for it eliminates the interaction between his genes and the mother's, the imbrication of the parents' differences in the *joint* act of procreation.
>
> The cloner does not beget himself: he sprouts from each of his genes' segments . . . in effect resolving all oedipal sexuality in favour of a "non-human" sex, a sex based on contingency and unmediated propagation.
>
> The subject, too, is gone, because identical duplication ends the division that constituted him . . . The only thing cloning enshrines . . . is the reiteration of the same: 1+1+1+1 etc. . . . Sex itself [is turned] into a useless function.
>
> (Baudrillard 1993, 115–16, emphasis in the original)

If cloning turns sex into a "useless function," that is because, for Baudrillard, sex's reproductive role is overridden by technology, its power to designate (sexual) difference as the foundation for humanity is rendered redundant, and the genetics of kinship are be replaced by artificial ties. Such condensations and elisions play on the slipperiness of the word "sex" and on the contested nature of the teleology of its power to signify (Butler 1990). Renaturalized here as the marker of subject formation, the privileging of heterosexuality (by implication) relegates all other forms of sexuality to the realm of sameness. Within this rhetorical logic, heterosexuality is life, and all else is death—or rather, heterosexuality guarantees mortality (and thus the necessary boundary between life and death), and all else threatens to bring an unbearable immortality (which is ultimately the end of life as we know it, just a living death by another name). If sex difference as the foundation for reproduction is lost, it seems, so too will be the subject and its relation to others. Sliding seamlessly from biological evolution to cultural formation and back again, Baudrillard elevates heterosexuality to the foundational cornerstone of civilization and casts its absence as the space of the inevitable and intolerable repetition of sameness, re-naturalizing kinship as "always already heterosexual" (Butler 2002) by combining the duration of history with the inevitability of evolutionary biology. As if in implicit dialogue with, or perhaps in defiance of, feminism (which has so thoroughly challenged such moves), Baudrillard rehearses a discursive logic which elides genes, sexual difference, heterosexuality, and reproduction. Here sex is the originary biological division upon which symbolic differences are then mapped, and upon which the symbolic order itself depends—rather than being the effect of a biological discourse whose reverse teleology makes it appear as origin.[9]

Since heterosexuality, according to Baudrillard, culturally anchors the biological progress of evolution in the symbolic order, premised on the biological necessity of reproducing sexual difference in the structure of reproduction itself, cloning is not alone in transgressing the laws of nature. Such a vision of heterosexuality as providing the necessary basis for evolutionary progress places all other modes of sexuality and reproduction outside both the natural and the symbolic orders. Sexual difference is the only difference that matters to reproduction, according to this rhetoric, and thus all other desires are merely a desire for the same. To designate sexual difference as the sole biological marker of a necessary psychic division for the subject is to place desire outside the heterosexual matrix, on the barren terrain of sameness (Stacey 1997). There is a paradox at the heart of this "paradigm of sameness" deployed throughout Baudrillard's writing on cloning which Edelman challenges so eloquently in his book *No Future: Queer Theory and the Death Drive* (2005). The trajectory that aligns sameness with death and then places both in opposition to a life-affirming heterosexual reproductive difference is one determined by the following paradox: "Homosexuality, though charged with, and convicted of, a future-negating sameness construed as reflecting its pathological inability to deal with the fact of difference," argues Edelman, "gets put in the position of difference from the heteronormativity," which despite its "persistent propaganda for its own propagation through sexual difference, refuses homosexuality's difference from the value of difference it claims as its own" (Edelman 2005, 60). Edelman quite rightly accuses Baudrillard of naturalizing heteronormativity in his celebration of "the triumph of sexed reproduction over genetic duplication" (ibid., 62). For Edelman, Baudrillard "sounds the note of futurism's persistent love song to itself, its fantasy of a dialectic capable of spinning meaning out of history and history out of desire" when he naturalizes the "trajectory from the replication he associates with genetic immortality to the procreation made possible by encountering sexual, and therefore genetic, difference" (ibid.). As we have seen, according to this logic, sexual sameness is a fatal threat not only to the social fabric and the symbolic order that regenerates it but also to the imperative to self-perpetuate that heterosexual difference guarantees. In short, to survive the threat of both genetic science and homosexuality, the species must, as Edelman puts it, "remain the same in its difference from the lethal sameness it condemns for its nullification of difference" (ibid., 63).

Whereas Baudrillard's 1993 essay on cloning provocatively renaturalizes heterosexuality without referencing the powerful critique of this logic from feminist and queer theorists, his later essay in 2000, though no less provocative, is at least in more explicit dialogue with the question of what he refers to as "sexual liberation." Here Baudrillard suggests that the only real sexual revolution was "the advent of sexuality in the evolution of living things, a duality that puts an end to perpetual indivision and successive iterations of the same" (Baudrillard 2000, 9). Provocatively placing the "sexual revolution" (evolution) side by side with the discourses of what he calls "sexual liberation" (the birth-control pill, feminism, and lesbian and gay politics), he suggests that although it may seem that the two have common goals, on closer inspection the "ambiguous repercussions" of the latter may be "completely opposed" to the progressive directions of the former: "first, sex was liberated from reproduction; today it is reproduction that is liberated from sex through asexual, biotechnological modes of reproduction such as artificial insemination or full body cloning" (ibid., 10). According to this teleological configuration, the desire for sexual liberation has apparently gone against the interests of evolutionary biology and sided (knowingly or not) with those scientists whose desire for cloning coincides with the blind selfishness of the sexual liberationists. If nature required women to accept their part in the mutual dependency of naturally evolved sex differences, then perhaps stepping outside this dyadic complementarity has threatened more than we realized; and if sexuality and reproduction were held together in the natural order of things for good evolutionary reasons, then perhaps the consequences of their separation (demanded by feminism) now coincide with bio-technology in reversing the most crucial underpinnings of the social order.

The dread of sameness haunting the genetic imaginary is here explicitly connected to fears of broader changes in the practices of sexuality and reproduction (even where these are never made explicit, they nevertheless pervade Baudrillard's writing on cloning). Thinking about changing definitions of life and death in the age of cloning necessarily requires a consideration of the ways in which multiple and overlapping conventional couplings are also being reconfigured: not only sexuality and reproduction but also kinship and parenthood, procreation and care, and male and female complementarity. Significantly, Baudrillard names artificial insemination alongside cloning in his designation of all biotechnical modes of reproduction as part of the problem. "Artificial insemination" is a very loose term, encompassing a broad range of practices from the low-tech self-insemination of donor sperm (favored by single women and lesbians) to the fertilization of the egg by the sperm outside the woman's body and its technical implantation in the womb (primarily used by heterosexual couples). Although artificial insemination is the named offense here, we might substitute any number of others which fall outside the traditional borders of the heterosexual family as the autonomous unit of natural procreation: in vitro fertilization, single parenting, lesbian and gay parenting, egg donation, and surrogacy. What Baudrillard designates as the deathly practices of sameness here are contrasted with the "life-giving" energies of heterosexuality as the foundation of culture (Butler 2002, 29). These anxieties about the proliferation of undesirable forms of sexual and reproductive sameness are driven by a heterosexual imperative which seeks to locate reproduction unequivocally within the supposedly biological regimes of sexual difference.

For Baudrillard, it is "sexual liberation" which has finally begun to undo the real sexual revolution (evolution) and now endangers the symbolic order itself with its irreverent rejection of the necessity of difference: "Sexual liberation, the so-called crowning achievement of the evolution of sexed forms of life, marks, in its full consequences, the end of the sexual revolution" (Baudrillard 2000, 11). Somehow, in their shared project of uncoupling sexuality and reproduction, science and feminism have removed sexual difference as the foundation of both biology and culture and have thus set humans adrift in a sea of unending sameness. According to Baudrillard, without its anchor in embodied sexual difference, sex becomes a "useless function" (Baudrillard 1993, 116, and 2000, 10). Both science and feminism have failed to anticipate the consequences of such ignorance and foolishness (and possibly selfishness): "The calculated benefits both of sexual liberation and of the scientific revolution are inextricably bound up with their negative countereffects" (Baudrillard 2000, 11). Feminism and genetic engineering are united in their mission to undo the basic foundations of nature and culture and to dislodge sexual difference as the cornerstone of civilization.

In the landscapes of the genetic imaginary, the proliferation of sexual and reproductive practices is seamlessly elided, condensed, and substituted one for the other: donor insemination becomes in vitrofertilization; egg donation becomes lesbian cloning; stem cell manufacture becomes designer babies. What lies at the heart of these moves is a profound anxiety about the destabilization of sexual difference as the cornerstone of culture and about the introduction of unnatural forms which separate sexuality, reproduction, procreation, and kinship. New techniques of reproductive and genetic intervention become inextricable from the detraditionalization of normative familial forms and modes of procreation. The threat to the place of sexual difference as the stabilizing force that unites biology and culture can be felt in both these new ways of inhabiting the social and sexual world and in these innovations in techno-scientific reproductive and genetic engineering.

In this account of the "phantasm underlying the whole genetic project" (Baudrillard 2002, 198) we see a renaturalization of the place of reproductive heterosexuality within the symbolic order. As Judith Butler has argued in a related context, the "symbolic order" is mobilized as corresponding to and ratifying an implied natural law (Butler 2002, 24). Here Butler is challenging the discourse operating in public debates about lesbian and gay partnership and adoption initiatives; she highlights the ways in which those who protest against such initiatives "proclaim that sexual difference is not only fundamental to culture, but to its transmissibility" (ibid., 32).[10] Through this essentialization of the symbolic order, sexual difference is conflated with cultural transmission. Here, heterosexuality is constructed as the "life-giving" foundation of culture, she argues, and homosexual parenting is a "practice that not only departs from nature and from culture, but centers on the dangerous and artificial fabrication of the human that is figured as a kind of violence or destruction" (ibid., 29). Just as in Baudrillard's rhetoric, the concepts of the "symbolic order" and "culture" are invoked here to "designate not the culturally variable formations of human life, but the universal conditions for human intelligibility" (ibid., 29).[11]

Similarly premised on the notion of sexual difference as the irreplaceable foundation of cultural transmission, sexual sameness in the genetic imaginary becomes intelligible only as the signifier of the end of both nature and culture. The genetic imaginary is haunted by the threat of a destructive and violent turn to sameness that robs culture of its reproductive motor, confirming the notion that only sexual difference connects the reproduction of life that flows between nature and culture. This rendition of the uncoupling of reproduction and sexuality assumes a new logic governed by sameness and, by extension, by a perverted sexuality organized around a fatal desire for the same. Conceptualizing the changing relations between biology, technology, and sexuality through the notion of sameness constitutes such changes as *necessarily* analogous. As Robyn Wiegman has argued, heteronormativity produces everything other-to-itself as a perverted desire for the same: "Queer desire . . . [is] many things, but the heteronormative idea that it was 'same sex' attraction or, worse, a mere mimicry of gender conformity, [is] absurd" (2006, 91). Such assumptions permeate a genetic imaginary that depends upon aligning sexual difference with the life-giving desires of reproductive heterosexuality and sexual sameness with atrophy and the perversion of the natural forces and teleologies of biology.

This chapter has mapped the forms of normativity which lie at the heart of the genetic imaginary. My reading of the expansive force of such an imaginary makes visible the elisions and conflations that enact its conceptual associations. Central to this force is the place of an assumed sexual sameness within the cultural terrain of technologized, nonreproductive sexuality. Distaste for what can be clustered together under the paradigm of sameness is affectively palpable. The horrors aligned in the dread of a perceived sexual, technological, and biological sameness works here as a conceptual conflation which carries the weight of the affective cultural disturbance of genetic engineering and cloning.

[2] She Is Not Herself

The Deviant Relations of *Alien: Resurrection*

Writing about the problematic question of spectatorial identification with the heroes of cloning films, Debbora Battaglia argues that "because the heroes here are multiple, not the autonomous egos of Freudian theory, we are *with* the owner and the owned at once, if we are with anyone; our subject-position identifies with a *relation*" (Battaglia 2001, 511).[1] Battagalia conceptualizes the figure of the clone in the cinema as a "supplement," or "something that supplies, or makes apparent, insufficiencies" (ibid., 496), for, "unlike the replicant which requires no connection to an original and is often seen questing for a connection, even a negative connection, to its makers (such as the toy maker and Tyrell Corporation CEO of *Blade Runner* [Ridley Scott, 1982])," she argues, "the *clone embodies the closest relation to the original*" (ibid., 506, emphasis added).[2] Taking my cue from Battaglia's important insights about the figure of the clone in the cinema, in this chapter I shall examine the embodiment of this relation in the new monsters of genetic engineering and cloning which confound the traditional boundaries of sameness and difference so central to the pleasures of the body-horror film. If the clone embodies a relation, as Battaglia contends, what is at stake if we consider the implications of this relation as one of sameness, and how are such relations constructed in the monstrous embodiment of science-fiction and horror films?

These questions emerge in dialogue with a number of feminist debates about the monstrosity of sameness. As Elizabeth Grosz has argued, monsters have traditionally been connected to the threat of duplication:

> Monsters involve all kinds of doubling of the human form, a duplication of the body or some of its parts. The major terata recognized throughout history are largely monsters of excess, with two or more heads, bodies, or limbs; [they may have] duplicated sexual organs . . . [There] is a horror at the possibility of our own imperfect duplication, a horror of submersion in an alien otherness, an incorporation in and by an other. (Grosz 1991, 36]

In what follows, I offer a reading of the configuration of cloning as the embodiment of what I call the relations of excessive sameness in *Alien: Resurrection* (Jean-Pierre Jeunet, 1997).[3] Whereas all cloning films could be read through the problem of too much sameness, insofar as they are concerned with technologies of copying (those within the science-fiction or body-horror genres probing the monstrous potentialities of such imitative experiments), I argue that the cloning of Ripley (Sigourney Weaver) in *Alien: Resurrection* is given cinematic life through a relay of anxieties connecting multiple and interrelated forms of biological, technological, and sexual sameness. What is at stake here is the extent and form of the visibility of sameness.

My reading of *Alien: Resurrection* extends a number of debates within feminist film theory about the transformations of the monstrously feminine, reproductive bodies of horror and science-fiction films into the threatening new iconographies of genetic engineering and cloning. To this effect, I put feminist theories of the geneticization of culture (Franklin, Lury, and Stacey, 2000; van Dijck 1998) and the teratological imagination (Braidotti 1994 and 2002, Stacey 1997) into dialogue with debates about the centrality of the abject for understanding the monstrous feminine in the cinema (Creed 1993). Building upon feminist challenges to Barbara Creed's theories of the abject with respect to the horrors of genetic engineering in *Alien: Resurrection*, I investigate how the queering of desire through the figure of the clone requires us to reconsider the problem of representing sameness in terms not only of duplicated embodiment (Constable 1999), but also of the film's self-conscious play with the performance of passing.

While the precarious boundary between monstrous and proper bodies pervades both the narrative structure and iconographic landscape of all the *Alien* films,[4] the fourth film in the series, *Alien: Resurrection*, places genetic engineering center stage by organizing its exploration of monstrousness around the story of the cloning of Ripley. Each *Alien* film involved the cinematic (re)incarnation of Sigourney Weaver as a different kind of Ripley: from action heroine in *Alien* (Ridley Scott, 1979) to marine cyborg in *Aliens* (James Cameron, 1986) and androgynous social outcast in *Alien 3* (David Fincher, 1992). The literal cloning of Ripley by scientists in *Alien: Resurrection* translates the problem of generic repetition into a scene of genetic enactment. But Ripley is not merely a clone of her former self. She is a transgenic clone: a clone whose original already combined the DNA of two different species, human and alien. Drawn from cellular traces of a Ripley impregnated by an alien (an act that led to her sacrificial suicide, as she plunged in a Christ-like pose from a great height down into the flames below, at the end of *Alien 3*), this latest version of Ripley combines her with the monster at the genetic level, producing the spectacle of transgenic kinship between the human and the alien. If the figure of the clone is generally constructed as the embodiment of a relation, as Battaglia argues, then Ripley as transgenic clone embodies the relation of original to copy and, simultaneously, the relation of human body to alien monster. As such, the story of Ripley's artificial reincarnation and transgenic kinship in *Alien: Resurrection* moves the series firmly into the spectacular realm of the scientific laboratory and the diseased, deformed, and mutating bodies resulting from mixing DNA that populate the spaces of the genetic imaginary.

Since the original version of *The Fly* (Kurt Neumann, 1958), perhaps the first film to explore the mixing of human and nonhuman molecules using technical experimentation, the mutational possibilities of alien DNA in the human body have generated bodies with spectacularly terrifying transformative potential. For Matthew Hills, the current cycle of "species-level biohorror" films draws on a long-standing preoccupation with the boundaries around the distinctiveness of the human in the horror genre, but seeks to "use gene discourse (not always successfully) within textual strategies of distinction and strategies of displayed intertextual cultural capital" (Hills 2005, 16). In body-horror films, the transgenic combinations of human DNA with DNA from outer space (*Alien: Resurrection* and *Species*) or from the animal kingdom (*The Fly* [David Cronenberg, 1986] and *The Island of Dr. Moreau* [John Frankenheimer, 1996]) and the experimentation or interference with animal DNA from the past or present (*Jurassic Park* [Steven Spielberg, 1993], *DNA* [William Mesa, 1997], and *Deep Blue Sea* [Renny Harlin, 1999]) result in disturbing hybrids, uncanny freaks, or unnaturally aggressive and threatening lifeforms. These fascinatingly repulsive genetic disturbances confuse the traditional dichotomies of nature/culture, human/nonhuman, and authentic/artificial: examples include the zombies of *The Eighteenth Angel* (William Bindley, 1998), the hybrids of human and alien DNA in *Alien: Resurrection* and *Species*, and the mutational genetic transformations of *Hulk* (Ang Lee, 2003).

Alien: Resurrection places the mutability of the cell at the heart of its spectacular display of monstrous bodies. The mesmerizing credit sequence shows cellular mutations in flowing motion: shot in close-up, the warm glow of the golden color; its reflective, shiny, sometimes sticky surface; and the repeated use of circular, interconnected patterns give a honeycomb look to the substance. But a much more sinister set of associations governs this sequence, as almost immediately cells mutate into more monstrous distortions. The waves of the flowing cells are accompanied by an eerie musical score that rises and falls to the same rhythmic pattern. A bodily form emerges as the moving contours begin to resemble mutating versions of skin, flesh, and bones. Crossing the usual internal-external boundary of bodily integrity, the slimy, mucoid substances and skeletal patterns are reminiscent of many of the reversals built into the design of the alien bodies in the previous films. Extreme close-up shots (accompanied by a distorting mirroring effect) combine with the fluidity of the constant mutation to thwart any attempt to fix the meaning or origin of the organs and body parts displayed, offering instead a sense of the endless transformational potentialities of cellular formation. Gradually, though, the forms become more recognizable as they become grotesquely distorted organs, ending

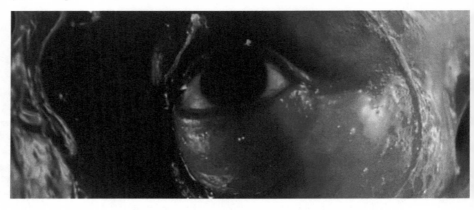

3. Body parts appear and disappear in the undulating cellular mass of
the opening credits of *Alien: Resurrection* (Jean-Pierre Jeunet, 1997).

the opening sequence with a kaleidoscope of monstrosities as teeth,
hair, eyes, and bones appear and disappear within the undulating fleshy
mass (see figure 3).

In *Metamorphoses: Towards a Materialist Theory of Becoming* (2002),
Rosi Braidotti argues that the fascination with the monstrous bodies
of scientific cell manipulation is inextricable from the new potential
for the mutability of the image associated with digitization and global
cybercultures. Braidotti's insistence on the significance of these technical intersectionalities in contemporary culture demands a consideration of the interface between genetic engineering and digital manipulation.[5] This broad challenge from Braidotti might be posed as a more
specific question here about the relation of cinematic and scientific
technologies, and of cloning and morphing in this film. In the opening sequence of *Alien: Resurrection*, for example, the continuous flow
of morphing cells and organs produces a sense of biological mutability
through perpetual visual transformation. As we watch this fluid movement of changing cell formations, the hidden monstrous organs and
features repeatedly surface and are reabsorbed, offering only a cautionary glimpse of the more dangerous and disturbing threats hidden in the
golden flow. This extended spectacle of cell mutation uses digital special effects to foreshadow the horrors of genetic engineering, a form of
scientific intervention into cellular life that threatens to produce monsters as well as marvels (see Stacey 1997).

According to Braidotti, the mutability represented by these respective techniques intersect and reinforce each other powerfully through their closely related monstrous associations. She writes of "a distinct teratological flair in contemporary cyberculture," in which "a proliferation of new monsters . . . often merely transpose into outer space very classical iconographic representations of monstrous others" (Braidotti 2002, 179). For her, "the monstrous or teratological imaginary expresses the social, cultural and symbolic mutations that are taking place around the phenomenon of techno-culture," and she suggests that "visual regimes of representation are at the heart of it" (ibid., 181). The current conjunction of genetic engineering with the cyberculture of globalization is given shape through what she calls the "teratological imaginary."

Teratology means both the scientific study of monsters and marvels and the study of teratomas, tumors which develop from an unfertilized germ cell (an egg or sperm) (see Stacey 1997). Teratomas that are cancers of the egg cell "have the capacity to differentiate into any of the specialized cell types of the adult body," and they are often "filled with clumps of matted hair, protruding lumps of bone, cartilage, bronchial and gastro-intestinal epithelium and even teeth" (Cooper 2004, 17).[6] These bizarre, parthenogenetic, disordered mixtures of tissue violate general cultural categories: they mix together life and death, health and illness, the normal and the pathological, the human and the monstrous.[7] This is the teratological imaginary, populated by the horrors evoked by such disturbing combinations of monstrous proportions. Since it includes spectacular images of cell misbehavior, the history of teratology presents a powerful iconographic repertoire for contemporary cultural representations of genetic manipulation.

The fascination with teratomas and excessive cell growth in *Alien: Resurrection* draws on the long-standing preoccupation in science-fiction films with monstrous bodies and on more recent fantasies about genetic engineering and cloning. The teratological imaginary conveys a renewed sense of fascination with deviant bodies at a time when scientific manipulation of cells might produce more terrifying distortions of the human form than nature does.[8] *Alien: Resurrection* produces an iconography of monstrous bodies in which medical science generates the spectacle of failed genetic offspring. In other words, biological deviance of the organic variety (disease) gives cinematic life to the terrifying monsters resulting from technological interference with biology (genetic engineering and cloning).

The significance of the teratological dimension of the spectacle of cellular mutations and distortions in the credit sequence is not fully elaborated until much later in the film, when those mutations are echoed in Ripley's encounter with the display of the monstrous, failed clones numbered one to seven. Following Ripley's successful cloning as a host body for the infant alien on the spaceship *Aurega* at the beginning of the film, her baby is removed from her chest and allowed to mature in a giant, transparent cylindrical container. The scientists responsible allow Ripley to remain alive out of curiosity. A pirate ship, *Betty*, arrives with illegal cargo (live bodies for hosts in which to breed the aliens) and a motley crew, including Call (Winona Ryder), a robot programmed for compassion who passes as human for most of the film, and who is trying to destroy the aliens (and Ripley, too, if necessary). Once the aliens begin to reproduce and destroy the humans, the remaining members of the crews of the spaceships combine and are led in flight by Ripley, repeating the narrative structure of the previous *Alien* films.

It is at this point that Ripley confronts her own status as the final outcome of a commodified series of genetic experiments. Delaying the remaining crew's flight, Ripley pauses to compare her own numerical tag (the number eight tattooed on her lower inside forearm) with the sequence one to seven on the entrance to the laboratory in which the clones are housed. Ripley's connection to her horrific predecessors is established through the numerical sequence, which, mimicking the technology which has produced the clones (DNA as genetic sequence), suggests Ripley's enslavement to the capitalist scientists who plan to make money from this experiment. The concentration-camp-style numbering on Ripley's arm confirms her status as an object of exchange in these capitalist relations—a "meat byproduct," as her creator calls her, with no rights to personhood. In this scene, the spectacle of the inhumanity of science, with the echoes of Holocaust eugenics, builds upon the connotations of slavery in the scene in which the scientist trains Ripley to be human—there, her hands and feet are in shackles to limit her violent and unpredictable outbursts, and she is presented as a "white slave" forced to conform through behavior modification. Following the much debated replicant film *Blade Runner*, in *Alien: Resurrection*, cyborg enslavement produces a historical reversal which appears to separate slavery from race: slavery is repositioned beyond race, as the political potential of the brutality of the combined ruthlessness of science and capitalism.[9] Created by a multiracial group of scientists, Ripley's whiteness cannot protect her from exploitation, manipulation, and objectification; white slavery becomes the ultimate horror in the fantasy of a genetically engineered future.

Part gallery, part laboratory—a kind of "unnatural history museum"— the scene of this scientific exhibit displays the spectacle of failed genetic experimentation like a chamber of horrors. A close-up shot from inside the room back through the glass, showing Ripley's face with the numbers one to seven in reverse across her forehead, indicates the distant trace of ancestral memory of her genetic kinship with these previous clones. The bodies of the first six dead clones are displayed in giant test tubes, suspended in a transparent solution, lit from above by a yellowish spotlight, and raised slightly above or slightly below eye level for human inspection. The golden color, combined with the repetition of the music from the credit sequence, returns us to the spectacle of the fascinatingly disturbing teratological grotesque with which the film began. Although very different in style, an aesthetic of fluidity links both scenes: in the credit sequence, the cellular forms mutate to reveal unexpectedly monstrous distortions; here, the continuous camera movement around the space both tracks and displays Ripley's trancelike journey through the monstrous history of her genetic ancestry. These transparent storage columns are spaced throughout the room, requiring Ripley to move among her predecessors as she slowly takes in the shocking visual evidence of her own prehistory. The shots alternate between medium shots and close-ups of the grotesque bodies of the clones, and long shots placing Ripley (and later Call) among them. Ripley's physical proximity to, and her touching of, the glass containers indicate her genetic kinship with and growing empathy for her teratological ancestry.

This gallery of genetically engineered monsters shows the spectacle of failed scientific recombinants of human and alien DNA. Embodiments of the threat of a new form of miscegenation, these transgenic clones are half-human and half-alien, and their hybrid status takes the visual shape of corporeal distortions. Recognizable traces of Ripley's hair, jawline, teeth, forehead, cheekbones, and nose are blended in the contours of the exaggerated facial features of others. The first six clones include fetal forms, tumorlike protrusions, misplaced organs, and distorted and extended features. The first resembles a fetus-shaped, curled-up old man whose spine continues into a tail, like a prehuman phase in evolutionary development (see figure 4). Another has elongated, witchlike fingers of alien proportion. Like the teratomas in the credit sequence, these monstrous births have inappropriate cell differentiation visible on their bodily surfaces: one has a spine or riblike bone formation running down the side of its leg; another has doubled organs—two mouths with enormous teeth. The attempt to interfere with normal cell division has produced excessive malignant growth: one clone has an extended skull doubling the size of its head, and another has a breastlike form growing out of its side. Ripley's body parts seem to have been disembodied and reassembled, echoing forms that tie her to the fate of her clone siblings. One of the dead clones on display has a mass of black hair like Ripley's, which floats Medusa-like in the liquid suspension—seen in a medium close-up shot prefiguring one of Ripley in a similar pose later in the film, during the underwater battle against the aliens (see figure 5). Shot from in front of the test tube through the yellow liquid, Ripley appears to accompany her dead relative inside the vessel. In their pathological deformities, the clones bring to the surface and make visible the deviant cellular composition hidden inside Ripley's own body.

The scene culminates in Ripley's encounter with the seventh clone, the most human of her ancestors. Bearing the traces of the previous failures (inappropriate bone formations penetrating the skin, a breast in the wrong place, elongated limbs), hooked up to machines that are keeping her alive and chained like a slave to her medical table, Number 7 is the ultimate cyborg outcome of the brutality of genetic science. Responding to Number 7's request to be killed, Ripley and Call torch the entire laboratory in an extended act of violent outrage. Cutting between shots from above, behind, and below, and using tilts to convey the distorting force of the attack, the spectacle shifts from the eerie stasis and containment of the science museum to a cathartic action sequence of elemental chaos as the golden glow of the fire combines with the reflective capacity of the liquid and glass. As Ripley backs out of the area, her panting breath connects her to Number 7, whom she has destroyed in an act that establishes her humanity through the affective connection of transgenic kinship.

In this display of unrestrained cell division, excessive organ growth, and deformed bodies defying the biological laws of "form, control,

4. Ripley (Sigourney Weaver) surrounded by failed cloning experiments
stored in giant test tubes in *Alien: Resurrection* (Jean-Pierre Jeunet, 1997).
5. Ripley (Sigourney Weaver) comes face to face with one of her
predecessors in *Alien: Resurrection* (Jean-Pierre Jeunet, 1997).

unity of design and function," *Alien: Resurrection* produces a fantasy
of genetic experimentation as the teratological grotesque (Varmus and
Weinberg 1993, 29). Like the malignant forms that are their prehisto-
ries, these failed clone bodies exhibit the fatal self-destruction of both
pathological and artificial cellular malfunction. The horrors of cloning,
of turning the culture of the copy into a biological possibility, are expressed through the display of excessive cell growth associated with
malignant disease, in which the body fails to recognize and expel the
monstrous proliferations within. Biological laws of cellular division are

defied in both cloning and malignant disease: the cells trick the body
into misrecognition; cellular similarity appears as difference, with fatal
consequences. The biological deviations of excessive cellular production of both teratomas and cloning combine in these monstrous figures
to show the threat of too much sameness in its most visible, material
form. Such disregard for the foundational requirements of difference at
the biological level is shown to have deadly effects in these failed clones.
The horrors evoked by such disordered teratological mixtures take on
an intensified force through the iconographies of techno-scientific experimentation with cell life. Moving reproduction out of the human
body and into the laboratory threatens to bring the monsters of nature
home to roost in the cloned bodies which are unnaturally, or too closely,
related to their original host. Ripley's connection to these monstrous
bodies, who are both her siblings and her predecessors, disrupts the
traditional linear generational teleologies of kinship.

Our conceptions of normal and pathological bodies and of our relatedness to our kin have been transformed by the possibilities of genetic
engineering and cloning, according to some critics.[10] As I discussed in
the introduction, cultural theorists—such as José van Dijck in *Imagenation: Popular Images of Genetics* (1998) and Dorothy Nelkin and Susan
Lindee in their study of the gene as a cultural icon—have claimed that
the so-called geneticization of culture, or "the genetic turn," means that
the genetic sciences (at the most general level) have redefined our understanding of the human body, reproduction, and disease (Nelkin and
Lindee 1995, 3). The genetic imaginary is populated by figures whose
new forms of kinship have been designed through the helping hand of
scientific intervention.[11] Fantasies of reproductive technologies, such
as in vitro fertilization, have pervaded popular culture in the form of a
technological fetishism, involving a disavowal of the mother's role, an
omnipotent fantasy of procreation without the mother that enables science, as Sarah Kember (1996) has argued, to fulfill the desire to father
itself. Cloning pushes such fantasies in new directions, placing the cell
as the unit of life at the heart of the troubling and yet fascinating potentialities of scientific experimentation. Cloning is only one dimension of
genetic engineering, but, according to Donna Haraway, it has a strikingly powerful grip on the popular imagination because it is "simultaneously a literal, a natural and a cultural technology, a science fiction

staple, and a mythic figure for the repetition of the same, for a stable identity and for a safe route through time seemingly outside human reach" (Haraway 1989, 368–69).

Cloning films arguably belong to the tradition of doubling that has pervaded the history of science-fiction films. As J. P. Telotte has argued, we might trace the continuities in the cinematic preoccupation with doubles and their threat to the individuality of the human form from *Frankenstein* (James Whale, 1931) through *Invasion of the Body Snatchers* (Don Siegel, 1956) to *Blade Runner*: "as exemplary texts, these films are commonly concerned with the human body as a double, and thus as an emblem of man's own blind space disconcertingly brought into contact with the specular" (Telotte 1990, 154). For Telotte, this preoccupation with the threat of the monstrous double is part of a general cultural anxiety about the other side of desire: "the horror genre typically conjures up 'monstrous' copies that, we would prefer to think, have no originals, no correspondence in our world. Their anomalous presence, however, fascinates us, even while it challenges our lexicon of everyday images" (ibid., 153). Since many science-fiction films are about the potential monstrosity of scientific and technological innovations in the name of the advancement of humanity, and since science-fiction films, as Mary Ann Doane (1990) has argued, repeatedly explore the relationship between the technological and the maternal (often showing the latter as contaminating the former), *Alien: Resurrection* conforms to its predecessors in both genres. Read through its inheritance, this cloning narrative rehearses the preoccupations of body-horror films more generally: these films are concerned with "the torture and agony of havoc wrought on a body devoid of control" (Brophy 1986, 10) and with "the ruination of the physical subject" (Boss 1986, 18). By exploring the potential monstrousness of genetic experimentation and cloning, *Alien: Resurrection* continues this evolutionary cycle of body-horror films. Following this trajectory, these cloned bodies might be symptomatic of contemporary anxieties about new life-forms and new forms of life, and about who will control their conception and reproduction.

This generic reading could be extended to confirm Creed's (1993) influential psychoanalytic theory of body-horror films as the ritual fantasy banishment of the abject, in which she reworks Kristeva's book on abjection, *Powers of Horror* (1982). As has been widely recapitulated, the process of abjection is the way we expel unwanted objects which remind us of our origins and our fate. These undesirable objects make us shudder, make our flesh creep, turn our stomachs; they have the capacity to simultaneously fascinate and revolt. According to Creed, it is the female body, particularly its reproductive capacity, which has been placed as the abject source of the narrative threat that must be resolved. The problem of separating from the maternal body is endlessly rehearsed in body-horror films, since, it is argued, we can never fully expel these reminders of the mutability of our boundaries, for the abject haunts the subject even after it has been expelled. Thus, Creed argues, the threat of being reabsorbed into the maternal body continues to preoccupy cultural fantasies and fictions: the womb represents the utmost in abjection. For example, in *Aliens*, according to Creed, the narrative strategies combine with a generic iconography of the womb to form the heart of the "monstrous feminine': "Woman's womb is viewed as horrifying [because] . . . it houses an alien life form, it causes alterations in the body, it leads to the act of birth . . . [The] womb . . . within patriarchal discourse . . . has been used to represent woman's body as marked, impure and a part of the natural/animal world" (Creed 1993, 49). Extending life beyond death, doubling the possibility of existence, and undermining the authenticity of individuality (the word "individual" originally meant indivisible), cloning threatens many of the foundations of contemporary culture, as I discussed in chapter 1. The clone might thus be the ultimate figure of abjection, if, as Kristeva argues, it is "not lack of cleanliness or health that causes abjection . . . but what disturbs identity, system, order. What does not respect borders, positions, rules. The in-between, the ambiguous, the composite" (Kristeva 1982, 4). Considering Ripley as a figure of genetic abjection, not only is she a clone, but she is one who was engineered already pregnant with an alien species: she is the literal embodiment of the full grotesque potentiality of the geneticized monstrous feminine. If, for Kristeva, the abject can never be properly banished and repeatedly returns to haunt the subject at the borders of its existence, Ripley literally embodies the constant threat of the abject. *Alien: Resurrection* produces its monstrous spectacles through an iconography that combines the clone, the monstrous birth, and the diseased female body. Creed's argument about body horror might thus be extended to an exploration of the genetic abject in this and other films, such as *Species* (Roger Donaldson, 1995), discussed in chapter 3.

Both generic and psychoanalytic theories here seem to point to the ways in which *Alien: Resurrection*, like its predecessors, rehearses the expulsion of the monstrous in order to reassure us of our bodily and individual integrity. In the light of these approaches, the scene in which Ripley destroys the monstrous seven clones, discussed above, could be read as a pivotal moment in the narrative drive toward humanizing Ripley, ridding her of a connection to the abject. In this scene, Number 7 is the abject figure who crosses the boundary between life and death, bringing death into life, and whose last breath calls Ripley to reimpose this distinction and kill her. She is Ripley's previous self, bearing an uncanny physical resemblance to her: too close to Ripley's own incarnation, Number 7 represents the problem of identity as excessive sameness. In destroying her, Ripley is, in part, destroying herself. Rehearsing her previous self-destruction at the end of *Alien 3*, Ripley revisits the narrative closure which failed to deliver its promise of total alien annihilation. In her destruction of the seven clones, who visualize and thereby literalize Ripley's deviant biology in their bodily evidence of cellular excesses, Ripley kills herself over and over again.

It is Ripley's and Call's emotional responses to Number 7's suffering that distinguishes them from the other, unfeeling humans in the crew. As Battaglia has pointed out, "it is not uncommon in films of human doubles that narratives of passing expose the insufficiency of dominant-culture originals" (Battaglia 2001, 509). In destroying her predecessors, Ripley demonstrates disgust for the excesses of genetic experimentation by eradicating the abject bodies it produced. In the process, she becomes more fully human herself by becoming an individual. Since the notion of being human depends in part upon the assertion of the singularity and originality of each human subject, in destroying the traces of her replication (the first seven clones), Ripley arguably begins to leave behind her cloned status. Paradoxically, by killing her clones out of a sense of compassion, she and Call become "more human than the humans"[12] on the ship. Her decision not to kill the human scientist responsible for creating this misery serves as a further sign of her humaneness, as well as her possession of the humanity which he lacks.

Despite the repeated display of the proliferation of sameness in the teratological monstrosities which give visual form so perfectly to the genetic abject, it has been argued that *Alien: Resurrection* undoes, rather than confirms, Creed's reading of the place of the monstrous feminine

in body horror. As Catherine Constable's important reading of the film
suggests, the narrative closure ultimately signals the failure to banish
the abject successfully, thus denying the spectator the pleasures of the
previous films in the series (Constable 1999). Moreover, the figure of
Ripley as transgenic clone confuses the conventional dichotomies that
the genre of body horror has historically relied upon. The problem with
the representation of Ripley as transgenic clone in *Alien: Resurrection*
is that she is both the threat and the hope, both human and alien, and
as such she can neither be destroyed, like her offspring and predecessors, nor can she straightforwardly be the heroine and the object of our
identification. Constable insists that since the clear boundary between
the normal and the monstrous body cannot be maintained in *Alien: Resurrection*, the continuing validity of Creed's suggestion that the horror
film is a ritual expulsion of the abject is called into question.

For Constable, *Alien: Resurrection* defies the psychoanalytic model
of the abject not only because of its narrative structure, but also because of its dependence upon the intersubjective models of embodiment throughout the film. For example, in the scene at the beginning
of the film in which the scientists surround the giant test tube which
displays the morphing figure that will be Ripley reborn as transgenic
clone, Ripley is connected backward and forward in time to her kin. She
is linked to the past through Newt, the child for whom she played surrogate mother in *Aliens* (the voice-over in the opening test-tube birth
scene in *Alien: Resurrection* repeats Newt's words from the second film
in the series), and to the future through the mother alien who will grow
up and reproduce, and to whom Ripley has just given birth—making
her both the monster's mother and sibling.[13] Constable's reading of the
complex dynamics of identity, memory, and temporality in this scene
demonstrates the limits of the psychoanalytic model of abjection for
interpreting the film's narrative trajectory:[14]

> Ripley's identity is thus set up as an intersection point. She is altered by
> giving birth to the queen just as the queen will later display the nature of
> Ripley's bequest to her. Within a traditional psychoanalytic model, these
> points of intersection would constitute a breakdown of the oppositional
> structures of identity. On this model, Ripley's new found memories would
> indicate a collapse of the division between the human and the monstrous,
> conflating Ripley with the alien queen. However, the beginning of *Alien:
> Resurrection* is complicated in that the alien's capacity for instinctual

memory also provides Ripley with a means for remembering Newt. The alien DNA is therefore reconfigured within Ripley to provide access to a specific relationship as well as to activate a species memory. The capacity for instinctual memory does not dissolve Ripley into the alien queen, but sets up a point of intersection between the distinct characters. (Ibid., 191)

For Constable, the transformation of Ripley's memory through her bodily mutation establishes a series of fluid exchanges between Ripley, Newt, and the alien queen, demonstrating the need for a new model of subjectivity based on interrelationality: "Theorizing the possibility of productive points of intersection between self and Other, human and monstrous, requires an entirely different model of subject formation. The morphing figure in the tube stands for the possibility of change through productive encounters with otherness" (ibid., 191).[15]

This argument is further reinforced if *Alien: Resurrection* is contrasted with its predecessors. Whereas in *Aliens*, Ripley and the queen meet in a battle between the good mother (human) and the bad mother (alien), in which they fight to the death to protect their offspring, in *Alien: Resurrection* the distinctions between human and nonhuman, human and animal, and human and machine have entirely collapsed. This is most vividly shown in two scenes. First, Ripley again abandons her flight from the ship with the crew and is drawn back to the queen's body as it writhes in pain, struggling to give birth. Ripley is viscerally called back, as if through an embodied memory of shared corporeality, and her movement becomes more like that of an insect or an animal in response to her sense of shared pain. Once hailed, Ripley is shown literally submerging herself in the viscous substance surrounding the queen. With an almost orgasmic abandonment of self, Ripley reconnects with the alien as if she is returning home to a familiar maternal body; as Claudia Springer argues, "by associating a deathlike loss of identity with sexuality, popular culture's cyborg imagery upholds a long-standing tradition of using loss of self as a metaphor for orgasm" (Springer 1996, 61). This incestuous sexual merger with the alien queen is followed by the birth of an alien biped who destroys the queen and pursues Ripley as she begins to regain her human consciousness and reconnect with the human endeavor to escape.

The second scene takes place in the escape shuttle, where Ripley has to expel the queen's (and thus her own) offspring, and, as Constable argues, "the way in which the monstrous child is dispatched by Ripley plays into the theme of intersecting identities that structure the film" (196). Ripley and the infant are shown in close-up, exchanging intimate caresses and mutual sniffing to convey a sense of love and loyalty through their shared nonhuman genes. This intimate mother-child reunion that "prefaces the infant's death," argues Constable, "means that the traditional dispatch of the final monster cannot be regarded as a triumph" (196) and contrasts strongly with the final destruction of the monsters in the first two films:

> In *Alien* Ripley dispatches the threat, securing the Symbolic space of the craft. In *Aliens* the battle between Ripley and the queen is a fight for species survival in which the human is pitted against the insectual. By *Alien Resurrection* the oppositional relation between the human and the inhuman has been completely reconfigured to form a series of intersecting potentialities. (Constable 1999, 197)

Constable's reading makes a convincing case that there is a problem in using the psychoanalytic theory of abjection to understand the body horror of *Alien: Resurrection*, insofar as Ripley literally embodies the impossibility of the requisite expulsion of the alien so vividly dramatized in the previous films in the series.

However, the film is preoccupied with the threat of excessive sameness and produces the desire for markers of difference, a boundaried self, and an individual body. As such, the film is constantly haunted by the abject, even as it demonstrates the problem of the expulsion of its monstrous manifestations. Rather than rejecting Creed's theory of abjection as inappropriate or outmoded, I suggest instead that there is an excessive presence of the abject in the repeated rehearsal of its failed expulsions in *Alien: Resurrection*. Reiterating the thwarted desire to banish the threat of the abject, the film plays with its genre's conventions and audience expectations. While the ultimate expulsion of the monster is indeed fraught with ambiguity, as Constable suggests, the impossibility of narrative closure around the alien banishment foregrounds this expectation precisely as a convention of the genre, bringing to the surface the formal moves of previous *Alien* films.

The film's self-conscious commentary on its own genre's conventions is immediately evident in the opening scene of the film discussed above. In some senses, this adult figure in vitro appears initially to be the "clean and proper body" (72) that Kristeva writes about: floating peacefully in a clear blue liquid, perfectly proportioned, clearly contoured, hairless, and unblemished, Ripley is the spectacle of the body as pure sculpted form. As Ripley is morphed into being on the screen, we are presented with the ultimate fantasy of autopoietic individuality: there is no visual evidence of attachment to the mother since the navel is almost invisible. This is the masculine fantasy of self-generation and autonomy that rids the subject of his haunting debt to nature (the maternal body), which Kristeva and Creed argue is rehearsed in the ritual of abjection. But the image of this perfect adult body morphing in vitro is accompanied by the uncanny voice-over: "my mommy said there were no monsters, no real ones anyway, but there are." Whereas Constable emphasizes the scene's significance in terms of cross-species memory and intersubjective embodiment, I stress here its playful reference to *Aliens* for audiences familiar with the series. For this moment juxtaposes the innocent, yet knowing, child's comment with the literal making of Ripley's body as a perfect image, deceptive in its surface perfection because Ripley is already pregnant with the alien and about to become the monster's mother. In appearing as the perfect clone, Ripley is thus transformed from her previous role as good mother of Newt— for whom she became a protective surrogate mother and risked her life in *Aliens*—to her new, if unwilling, status as mother of the bad mother, the alien queen. In other words, the voice-over takes on an increasingly ironic status as the alien is shown to be not simply connected to this perfectly formed body, when the scientists perform a cesarean section to remove the fetal monster from Ripley's chest, but to be inextricable from her body and her psyche as she repeatedly demonstrates that it still resides within her. The pleasing fantasy of perfect individuality suggested by the adult fetus in vitro at the beginning of the scene is thus not only undone by the revelation that Ripley is the apparently human by-product of a barbaric scientific experiment, but is also playfully reversed in that Ripley is a transgenic clone, designed by scientists as the host mother for her alien infant. Defying the logic of individuality and singularity by combining two species and by being a clone, Ripley is both herself and not herself.

This generic self-referentiality is present throughout the narrative in its witty citation of the shots, dialogue, and mise-en-scène of its predecessor films. Such self-commentary has important implications for how we read the problematization of the abject in relation to the film's body horror. Considered as a form of knowing reiteration of the genre's conventions, not only does the figure of the transgenic clone confound the fantasy of any permanent expulsion of the abject other, but, in staging the impossibility of such an expulsion, Ripley can be read as repeatedly enacting the compulsive repetition of abjection itself, thus revealing the form of its foundational mechanisms. As Judith Butler has argued in relation to the regulatory functions of essentializing gender norms more generally: "When the disorganization and disaggregation of the field of bodies disrupt the regulatory fiction of heterosexual coherence, it seems that the expressive model loses its descriptive force. That regulatory ideal is then exposed as a norm and a fiction that disguises itself as a developmental law regulating the sexual field that it purports to describe" (Butler 1990, 136). Following Butler's claim, we might argue that *Alien: Resurrection* threatens the power of the theory of abjection, exposing not only the potential loss of its descriptive force but also the regulatory function which lies at the heart of its theoretical foundations. In this vein, as transgenic clone, Ripley's rehearsal of the problem of the me-not me boundary could be read as demonstrating its normative imperative but refusing its reassuring reinstatement (however temporary this might be): she remains both human and alien, both organic and synthetic, both Ripley and not-Ripley. When confronted by Call with the question "Are you Ripley?" early in the film, she replies, "I am not her." A technological deviation from the biological norms of reproduction, Ripley embodies the impossibility of the expulsion of the abject, exposing the regulatory principles in body-horror film. The structures of abjection, which offered such a powerful account of the appeal of previous *Alien* films, are pushed to the surface precisely through the failure to instantiate the boundaries around the human subject in *Alien: Resurrection*. Butler's challenge to the normative function of theories of the abject in gender-identity formation looks at the question of the destabilization of the subject through the loss of a fantasy of internal coherence. She writes:

> "Inner" and "outer" make sense only with reference to a mediating boundary that strives for stability. And this stability, this coherence is determined in large part by cultural orders that sanction the subject and compel its differentiation from the abject. Hence "inner" and "outer" constitute a binary distinction that stabilizes and consolidates the coherent subject.

When the subject is challenged, the meaning and necessity of the terms are subject to displacement. If the "inner world" no longer designates a topos, then the internal fixity of the self and indeed, the internal locale of gender identity, become similarly suspect. (Ibid., 134)

How are the suspect relations of sameness that are hidden in Ripley's body displaced onto the surfaces of her monstrous siblings on display? What happens when Ripley's "inner world" as transgenic clone "no longer designates" a recognizable "topos"? Extending Butler's argument to look at the "redescription" of intrapsychic processes, such as abjection, in terms of the "surface politics of the body" or, rather, in terms of the encounter between the surfaces of two bodies, such as Ripley's and Call's, we might ask: to what extent do the dynamics between them externalize this problem of the impossible conflicts of transgenic replication within Ripley's cloned body (ibid., 135)? In the final two sections of this chapter, I examine how Ripley's hidden genetic depth as the embodiment of excessive sameness is given form through her relationship with Call—the robotic double of Ripley's younger incarnations. These dynamics between Ripley and Call make visible Ripley's suspect internal topos, and it is here that the embodiment of the relations of sameness are most fully and often playfully elaborated.

It is not only Ripley's shifting status as clone, or her problematic genealogy as displayed in the teratological spectacle, which cast her as deviant in *Alien: Resurrection*; it is also her association with queer desires. This association of biological sameness with sexual sameness is a common trope in a number of films in which duplication is positioned at the heart of the most fundamental structure of the human biology (DNA). Anxieties about the detraditionalization of reproduction are closely aligned here to fears about new modes of sexual identification associated with a narcissistically designated homoeroticism. Films as different as *Alien: Resurrection, Evolution* (Ivan Reitman, 2001), *Gattaca, Multiplicity, The 6th Day* (Roger Spottiswoode, 2000), *Teknolust*, and *The Twilight of the Golds* (Ross Kagan Marks, 1997) all connect homoerotic desires with biological duplication, albeit through very different genetic and cinematic strategies. *Alien: Resurrection*'s play upon the trope of excessive sameness stretches across biological malignancies, genetic replications, and homoerotic intimacies; in *Multiplicity* the cloned husband becomes increasingly camp the further he gets from his original masculine body; in *The 6th Day* the protagonist's narcissistic aggression (because he has to share his wife sexually with his clone) transforms heterosexual transgression into the possibility of homosexual threat; and *Teknolust* designates two of the cloned female triplets a same-sex couple. This distinctly queer dimension to these films associates homo-erotic desires and homosocial affiliations with unnatural biologies.

What Constable passes over in her analysis is the extent to which Ripley's mutable biological identity is established through the homo-eroticism of her interrelationality with Call in the film, which is established extra-diegetically (playing upon Sigourney Weaver's star image as lesbian icon) as well as diegetically. Thus, while Constable is right to insist upon the confusion of the categories of subject, object, and abject in psychoanalytic theory in producing a close reading of Ripley's monstrous maternity, she ignores how this is further complicated by another structuring relation: the heterosexual-homosexual distinction. It is precisely Ripley's multiplicity as transgenic clone and the mother of another species that is deployed to establish a homoerotic dynamic with Call.

Celebrated as the androgynous action heroine who defied traditional Hollywood conventions of feminine passivity, Weaver as a star became a lesbian icon during the 1980s and 1990s. Through her participation in violent action films and her characters' use of heavy weaponry and technological know-how, Weaver's image was associated with unusual physical strength and stature and a heroic agency exceptional for a woman in Hollywood films.[16] Keen to separate such gender trouble from any lesbian connotation, publicity around Weaver's star image makes much of her femininity and heterosexuality off-screen, as in the following article from *Film Review*:

> Appearances are deceiving. The classically trained actress is softer and more delicate off-screen than the characters she has played. She is no aloof and self-assured woman with an air of command. "I have come to embody feminine strength and confidence, but in real life my movie portrayals reveal little of how I feel about myself . . . It was an odd thing being thought of as the female Harrison Ford because I am not a fan of scary movies. In fact, I am very squeamish about horror films. I miss half of them because I look away at all the scary parts. My husband has to tell me when it's okay to look at the screen again." (Weaver 1997, 35)

6. Ripley (Sigourney Weaver) as action heroine in
Alien: Resurrection (Jean-Pierre Jeunet, 1997).

In contrast to the actress's self-proclaimed squeamishness, in *Alien: Resurrection* Ripley is almost parodically encoded for strength and power, with her curly black hair swept back, her padded vest, her tense body, and her "just you try it" look: she is a kind of heavy-metal, female Braveheart, with a touch of the mutability and artificiality of Michael Jackson in his pop video *Thriller* (John Landis, 1983; see figure 6).[17] Weaver's status as popular lesbian icon is given diegetic elaboration in *Alien: Resurrection* through Ripley and Call's eroticized dynamic.[18] Their androgyny is echoed in their shared status as scientific inventions, which sets them apart from the human crew and produces an intense intimacy between them. This plays not only upon Weaver's status as a lesbian icon, but also upon associations between cloning and homosexuality: from the colloquialism for a particular gay male cultural style (the clone) to the more general assumption that homosexual desire is inextricable from narcissism, understood as a desire for oneself (see chapters 1 and 4).

In the light of these broader cultural associations, we might read Ripley as cloned in more ways than one in the film: she is both a genetic clone herself and is brought face to face with a previous incarnation of herself in the figure of Call, who represents a visual enactment of Ripley's hidden biological depths. Call is portrayed by Winona Ryder, a youthful star who both resembles a younger Weaver and reminds us in this film of Ripley's earlier incarnations in the series. Stories about the dynamic between the two female stars in popular circulation contribute to their positions along the axis of similarity and imitation. In the article from *Film Review* cited above, Weaver is described as being pleased at having been Ryder's role model. Having firmly established Weaver's heterosexuality, the strength of an attachment between the older and younger stars can be introduced through the discourse of the schoolgirl crush: Weaver "is proud of her action icon status and touched when . . . [her] co-star Winona Ryder told her that Ripley was her hero. 'Winona told me that all through high school, she had slept under a poster of me in *Alien*. I was touched'" (ibid., 33).

This dynamic of resemblance, admiration, and feminine mimicry drives the relations between Ripley and Call throughout the film. In the scene where they first meet, soon after Call has entered the *Aurega* with the pirate crew, their visual mirroring provides an external analogy to Ripley's cloned embodiment. Call (still passing as human) arrives at Ripley's door dressed in a dark-gray combat suit similar to the one worn by Ripley in the earlier *Alien* films; she has blanched white skin; dark, almost black, hair; large eyes; and a petite nose: she is Ripley's successor and also resembles a version of Ripley's past self. The hyperwhiteness shared by the two characters connects them to each other and to an ex- aggerated ideal of physical perfection: their similar bone structure (high cheekbones) and proportioned features (big eyes, small noses, and wide mouths) echo each other in their conformity to cultural norms of white feminine beauty. Their skin appears at times almost translucent (Ripley lies like a waxwork model on the laboratory table; Call appears like a ghostly presence returned from the dead) as their whiteness becomes evidence of their automated femininities and their synthetic origins.

The homoerotic connotations of this doubling are staged in the ex- plicitly sexualized dynamic of their first encounter. In this scene, Ripley is the feline dominatrix, playing with her prey, as Call struggles to hold her own with the older and wiser cyborg. The encounter is structured like a sadomasochistic seduction scene, in which the two alternate in positions of power and submission. Secretly entering Ripley's guarded panoptic cell, Call is seen in a close-up, low-angle shot from behind that shows her slowly drawing a knife from inside one of her steel-trimmed leather boots; she thus has a towering physical advantage over the su- pine, and apparently sleeping, Ripley, who is stretched out in front of her on the floor. Lying on her back with one arm placed above her head in the traditionally feminine position of submission, Ripley appears ready for seduction, if not penetration. She is dressed in a tight, sleeveless, dark leather waistcoat whose front lacing has been left open down to her breast level. As Call lifts the opening of the waistcoat with her knife, revealing the long scar between Ripley's breasts (from her cesarean delivery of the alien from her chest), Ripley's posture is one of utter exposure and vulnerability. The sexual signifiers proliferate: Call has found the entrance to Ripley's body that she sought and makes it visible with her phallic instrument. But as Call inspects Ripley, one deception (Call's entry into the cell with false identity papers) is superseded by another—Ripley is feigning sleep, and moreover, not fearful of death at all, she invites Call to "finish the job," to penetrate her with the knife and kill her. Taking control of the exchange, Ripley moves into a feline pouncing position and plays with Call's curiosity flirtatiously. In a two-shot, the women kneel facing each other, showing their almost identical profiles in mirroring symmetry. As Ripley replies with her original military title and number, she takes hold of the knife and forces Call to penetrate her hand with it (thus finishing the sexual act Call began). This violent gesture produces no blood or pain and offers a double answer to the question of her true identity: she both is and is not herself (see figure 7). Once penetrated by the knife, Ripley cedes power to Call, who reiterates that she is not Ripley. As the knife is slowly withdrawn from Ripley's hand, she gazes intensely at her younger mimic in close-up. Ripley takes Call's hand and traces her facial contours with it, offering tactile evidence of her instinctive feeling that the alien is still part of her. The homoerotic physical intimacy of this exchange culminates in a final move: straightening up on her knees, towering over her tiny intruder, Ripley encircles Call's face with long fingers and clawlike nails painted black. Half caressing, half threatening, Ripley reverses Call's touch (see figure 8). Switching back into dominatrix mode, Ripley grabs Call by the throat, holding her face roughly in her large hands, looming over her in a conventional heterosexual pose that is prelude to a first kiss. Call submits once again to Ripley's superior strength and is inspected, teased, and violently cast aside when Ripley is done.

This scene establishes an overtly homoerotic dynamic to Ripley and Call's relationship: the emphasis on touching, looking, role play, and intimate confessional dialogue combines with the status of Weaver as lesbian icon to produce a highly sexualized encounter. The use of close-up shots of Call's and Ripley's faces as they share their half-whispered

7. Call (Winona Ryder) penetrates the hand of Ripley (Sigourney
Weaver) with a knife in *Alien: Resurrection* (Jean-Pierre Jeunet, 1997).
8. Ripley (Sigourney Weaver) embraces Call (Winona Ryder) in *Alien:
Resurrection* (Jean-Pierre Jeunet, 1997).

secret knowledge confers an erotic intensity to their exchange. The
physical resemblance across the generations between the two actors
introduces a narcissistic feel to their intimacy and constructs homoeroticism through an aesthetic of sameness and repetition. The scene
establishes associations between physical resemblance and same-sex
desire, and between both of these and cloning.[19] Such an emphasis on
the visual mirroring of Ripley and Call places lesbian desire within a
mise-en-scène of excessive sameness: their exchanges throughout the
film emphasize their similarity, narcissistic recognition, sibling rivalry,
and even regressive mother love. Thus, the problem posed by Ripley's

biological deviance as transgenic clone is given surface presence in her encounter with Call: she is faced with a vision of her own abject duplicity. But these more abject overtones of the unnatural or cloying desire for sameness rather than sexual difference are arguably undercut by the film's self-conscious use of multiple associations of cultural codes of passing, in which sexual identity and synthetic origins are placed in ironic dialogue with each other.

Ripley and Call are both cyborgs (one a transgenic clone, the other a robot) who can pass as human, and each has to come out as nonhuman at different points in the film. They share knowledge, emotions, and recognition of their outsider status in relation to the values of masculine science and technology (even though they were created by it). There is even a structural echoing of Ripley's coming-out scene (discussed above) with the later one in which Call is forced to come out as a robot. In both scenes, the sexual connotations are explicit: as Springer has argued, "collapsing the boundary between what is human and what is technological is often represented as a sexual act in popular culture" (Springer 1996, 61). The second of their shared scenes reiterates the sexual encoding of their intimate exposure of each other's wounds and the penetration of each other's bodies through an exchange of touch in the first scene. Like Ripley, Call comes back from the dead when she rejoins the crew having been fatally wounded in the chest. Just as Call had earlier forced Ripley to confront her cyborg status, now Ripley reverses the demand. In a gesture that refers directly back to Call's, when she opened Ripley's clothing to reveal her cesarean scar, Ripley slowly and gently peels back Call's jacket to expose a wound in her side, oozing sticky white fluid. Echoing the sexual connotations of the guided penetration of Ripley's hand by Call's knife at their first meeting, Ripley slides her fingers into the cyborg's wound and rubs the viscous milky substance between them as she exits. Just as the status of Ripley's body as nonhuman was established when Call's knife entered it and neither blood nor pain resulted, so Call's robotic status is first indicated through the penetration of her body by Ripley's hand.[20] Throughout this scene, Call's lowered eyes, silence, and awkwardness, and finally tears all indicate the pain of hidden secrets and the shame of the deception involved in passing.

Call's exposure as a robot is rendered visible in a scene that replicates the revelation that Ripley is not herself, and this mirroring is amplified through the particular style of Call's robotic design. She is not only a robot but an "auton"—a "second generation" who should have been destroyed. Like Ripley, who is not a straightforward clone but a transgenic one, combining two species, Call is not a straightforward robot, designed by humans: she is a robot designed by robots. The cyborgs share common genealogical patterns: they embody a doubled duplication and have a special relationship to their nonhuman ancestors (Ripley to the alien, Call to computers); they have been manufactured by forces beyond their control (science and capitalism); and they should be dead (Ripley was forcibly reborn by scientists, and Call escaped the destruction of all autons by the government).

Where the associations of homoeroticism with an unnatural attachment to sameness might appear to confirm the normative drive of abjection in *Alien: Resurrection*, the reconfiguration of these relations of sameness as synthetic function pushes such normativity to the surface. The film repeatedly presents the problem of the threat of the abject and then translates it into a technological configuration that displays its formal conventions. Homoerotic intensity is translated into a more general intimacy of shared posthuman embodiment, in which Ripley's and Call's status as transgenic clone and robot, respectively, connect them through a corporeal and emotional empathy.[21] This is most clearly established at the moment when Call is transformed into the ship's computer in the second of her one-on-one scenes with Ripley, in the privacy of the ship's chapel. Call's connection to the mainframe is possible through a cable hidden in the Bible in the chapel (religion now functioning only to serve science); Call extracts a sinewy thread from the port in her forearm and inserts the computer lead like an intravenous needle. After fishing around for a connection, she begins to speak to the computer, named Father, giving information on the ship's power loss. As Call speaks computer commands with her new electronic voice, she enacts what Katherine Hayles has defined as the posthuman circulation of information "among different material substrates" (Hayles 1999, 1). The separation of information from bodies that is performed here reiterates a parallel disjuncture in Ripley's evolution discussed earlier: the dislocation of memories from subjects (Ripley has memories beyond the lived experience of her previous incarnations). As Michael Lundin has argued, both characters are thus placed within the realm of the posthuman, where, according to Hayles, "there are no essential differences or absolute demarcations between bodily existence and computer simulation, cybernetic mechanism and biological organism, robot teleology and human goals" (ibid., 3).

The transformation of Call in this scene places her alongside Ripley within a shared frame of posthuman mutability: "the posthuman subject is an amalgam, a collection of heterogeneous components, a material-informational entity whose boundaries undergo continuous construction and reconstruction" (ibid.). In a rapidly edited sequence of shots of sliding doors being released, and lights and computers being activated, the camera's roaming through the spaceships gives Call's posthuman status a kind of ubiquitous presence: with her full capacity of technological convergence, Call can play God. The combination of Call's cyber-knowledge and Ripley's instinctive judgment allows the cyborgs to control the ship and the shuttle. With their collaborative power, Ripley and Call together seem invincible. The homoerotic intimacy between them in this scene, established through their increasing knowledge of each other's posthuman technological potentialities, extends beyond their bodies into the realm of the interface itself. Their erotic exchange combines the intimacy of their shared embodied recognition with "erotic interfacing," which, Springer argues, "is, after all, purely mental and non-physical; it theoretically allows a free play of imagination" (Springer 1996, 68). *Alien: Resurrection* thus redefines homoeroticism as a kind of sexual interface in which the characters' knowledge of the techniques of replication and mutation produce their repeated sameness as artificiality and sexual deviance.

The cloning of Ripley in *Alien: Resurrection* produces a horror of and fascination with seeing the reverse logic of what heterosexuality was perceived to guarantee: sexual difference, sexual reproduction, and embodied maternity and paternity. To disorder these couplings is to threaten the foundations that seemed to ensure the compatibility of biological and cultural reproduction: with the threat to the integrity of individuality comes a threat to the logics of heterosexuality within this queer, posthuman landscape. As the transgenic clone whose pregnancy continues to connect her to an alien species even after she has given birth, and as the lesbian icon who plays a homoerotic power game with the newly arrived robotic version of her younger self, Ripley is constantly shown becoming someone else through her connections to

9. Ripley (Sigourney Weaver) and Call (Winona Ryder) in rhymed profile in *Alien: Resurrection* (Jean-Pierre Jeunet, 1997).

other bodies. In defining herself against the singularity of the Ripley of previous *Alien* films ("I am not her"), this resurrected Ripley has an indeterminate status: she is the not-Ripley or the more-than-Ripley. Embodying the informationalization of "life itself,"[22] Ripley's shifting identifications visualize genetic engineering and literalize its biological forms of mimicry. Her self-replication problematizes traditional categories of classification dividing the human and the nonhuman, the organic and the inorganic.[23] In configuring excessive sameness across pathological, technological, and sexual domains, *Alien: Resurrection* relentlessly demonstrates the impossibility of the expulsion of the abject other in a geneticized future. The consequent excessive presence of the abject in the film denies viewers the generic pleasures of its permanent expulsion, rehearsing and exposing the normativity at the heart of rituals of abjection.

Alien: Resurrection pushes the conventions of its genre to such limits that it produces excesses that are hard to contain within its forms of closure without becoming an ironic commentary upon the formalities of its predecessors. If the cloning of Ripley, pregnant with an alien, generates narrative problems for the traditional expulsion of the abject outlined by Creed, then the extension of the genre with this fourth film presents similar problems of duplication. Difference in repetition becomes both a biological and a cultural problem. As Constable argues, the film fails to reestablish the traditional boundaries around the human body to secure its integrity in the future. The biological, technological, and sexual forms of deviance in *Alien: Resurrection* are all presented as signs of too much sameness: teratomas, clones, and homoeroticism. Authenticity and individuality are called into question, not only by the threat posed by these forms of sameness, but also by the generic self-referentiality of the *Alien* series. As such, *Alien: Resurrection* rehearses the cultural analogy of the biological form of reproduction it explores. The audience is offered none of the conventional guarantees or reassurances of the body-horror genre. Instead Ripley and Call, two cyborgs, save the earth and are the signs of hope for an unknown future. Shown in *Titanic*-like double profile (see figure 9), they brace themselves against the impact with the earth's atmosphere and look ahead to the bright new future, signaled through a series of clichéd cloudscape shots. The survivors who accompany them are those whose bodies are coded as the most deviant of the pirate crew: the high-tech weapons expert in the wheelchair, Vreiss (Dominic Pinon), and the criminal thug, Johner (Ron Perlman), whose ruthless exploitation of others has no limits. The future belongs to the deviants—criminals, posthumans, and cyborgs.

[3]   Screening the Gene

Femininity as Code in *Species*

In a short essay titled "Sameness Is All," the psychoanalyst and cultural critic Adam Phillips tells of a conceptual confusion that became evident in one of his clinical cases. A child had been referred to him for school phobia, which started a year after her younger sister was born. Phillips offers the following account of their therapeutic exchange:

> In her second session [she told me] that when she grew up she was "going to do clothing." I said, "Make clothes for people?" and she said, "No, no, clothing . . . you know, when you make everyone wear the same uniform, like the headmistress does . . . we learned about it in biology." I said, "If everyone wears the same uniform, no one's special." She thought about this for a bit and then said, "Yes, no one's special but everyone's safe." I was thinking then, though couldn't find a way of saying it, that if everyone was the same there would be no envy; but she interrupted my thoughts by saying, "The teacher told us that when you do clothing you don't need a mummy and a daddy, you just need a scientist. A man . . . it's like twins. All the babies are the same." . . . I said, "If your sister was exactly the same

as you, maybe you could go to school," and she said "Yes," with some rel-
ish, "I could be at home and school at the same time . . . everything!"
(Phillips 1998, 89–90)

This extraordinary story of confusion between cloning and clothing
encapsulates perfectly a tension I wish to explore in this chapter: the
discrepancy between biological and cultural design, between what science can guarantee and what outward appearance might achieve, between the reproduction of embodied identity and the reproduction of
an image. This child's fantasy offers a resolution to the problem of sibling rivalry and parental control by positing a male scientist at the head
of a system in which similarity guarantees safety (the absence of envy)
and duplication overcomes the problem of displacement by a younger
sibling (simultaneous presence in two different places): it seems to allow the child to have "everything" she wants all at once, to fulfill all her
wishes simultaneously. The threat of reproducing sameness genetically
(cloning) is transposed onto the production of culturally uniformed
sameness (clothing).

The conflation of biological and cultural design in this story raises
the question of how identities are encoded and decoded. This in turn
points to what we might in shorthand refer to as the dialogics of identity: the ways in which identities are not simply constituted, embodied,
or lived out by subjects themselves, but are also always read by others.[1]
To say this, however, does not imply that they will be correctly read,
or that such a thing as a correct reading is even possible (Butler 2005).
Subjects do not intentionally encode their identities in ways which are
then decoded correctly or not by external others; rather, identities are
produced through a complex set of practices through which subjects
are constituted and in turn constitute each other. This dialogic process of identity production requires multiple cultural competencies to
achieve an interpretative exchange (or failure thereof) between subjects,
or among subjects and institutions. Indeed, the constitution of identity
is precisely an exchange between different subjects involving a complex nexus of recognitions and misrecognitions, revelations and occlusions, and expressions and impersonations. Identity is thus not a personal possession, or an individually authored script; it is a collaborative
achievement that is tentative, sometimes transitory, and often fragile.

If we accept that identity is a process rather than a product, a dialogic
exchange rather than a personal possession, then current processes of

geneticization complicate this picture in some interesting ways. On the one hand, a technique like genetic screening appears to offer an unchanging identity in the postmodern world of insecurity, instability, and identity fraud. As biological and geographical controls merge, DNA screening and fingerprinting promise definitive border control and rapid detection of crimes. In reproductive matters, preimplantation diagnosis and embryo screening generate fantasies of predictability and control. Genealogies now finally seem to be scientifically quantifiable and demonstrable; pedigrees now seem incontestable. In both legal and scientific terms, DNA testing has become the means to ascertain the truth of a person's identity that supposedly precludes all doubt. But, on the other hand, these genetic techniques have also introduced new insecurities: if life can now be manufactured, wherein lies its authenticity? If, as Katherine Hayles cautions, in the posthuman age *"information [has] lost its body,"* what new forms of cultural competence are needed to make it intelligible? And if she is correct to argue in respect to cybernetics and informatics that "a defining characteristic of the present cultural moment is the belief that information can circulate unchanged among different material substrates," what is the significance of that claim for geneticized bodies (Hayles 1999, 1–2, emphasis in original)? Put simply, how is the notion of cultural intelligibility transformed through new modes of biological legibility in the genetic imaginary?

This chapter examines the anxieties surrounding the reconfiguration of the boundaries around the body of the woman through the transferability of its informational components and the reproductive, imitative potentialities of genetic engineering and cloning. In the first half of the chapter, I discuss the implications of the emergence of the notion of the gene as a code; in the second half, I move on to consider the film *Species* (Roger Donaldson, 1995) through these debates about genetic encoding. Overall, this chapter examines how the woman's body as an investigative enigma in the cinema is transformed by the notion of the genetic code in such a way as to literalize the metaphor upon which the history of its imaginative status is premised. Tracing the notion of encoding femininity from the laboratory to consumer culture and back again, it interrogates the desire to control the woman's body and its reproductive capacities through a fantasy of the body's legibility as information that might be read correctly (and hence manipulated and controlled), if only its genetic depths and secret codes could be made visible.

I begin by raising two related issues: first, the gene cannot be captured as a photograph; and second, the gene has been popularly conceptualized through a proliferation of linguistic tropes.[2] Unlike the fetus and the cell, no image of the gene has taken on iconic status in contemporary culture (Franklin, Lury, and Stacey 2000, 19–44). In the absence of such an iconic image, genetic discourse has proliferated multiple metaphors through which the gene has been produced as imaginable, knowable, and tangible. We might ask: if the gene is not readily visualizable, how is it legible, and what does it mean to read someone's genes? In her elegant account of the "metaphors of twentieth-century biology," *Refiguring Life*, Evelyn Fox Keller tracks the shifts in metaphors for the gene during the twentieth century, including the recent radical transformation of biology by computer science:

> It is not only that we now have different ways of talking of the body (for example, as a computer, an information-processing network, or a multiple input-multiple output transducer) but that, because of the advent of the modern computer (and other new technologies), we now have dramatically new ways of experiencing and interacting with that body . . . it has already been constitutively transformed. (Keller 1995, xvii–xviii)

For Keller, practices, knowledge, and metaphors operate in and through each other in such a way that the language of biology is not only embedded within culture, but also transforms it in the most material ways. Typically, these metaphors have cast genetics as a language which holds the secrets of our bodily information. José van Dijck argues that with the discovery of DNA's structure by Watson and Crick in 1953, genetics has been given lay meaning through the metaphor of language: "the four bases of DNA . . . are labelled as the letters in the alphabet of DNA, and the endless combinations of letters constitute 'sentences' in a 'book,'" which then combine to "form the instructions for our genetic make-up, and knowledge of this structure may lead to control of heredity" (van Dijck 1998, 36). Although this tendency is not entirely new, the "legibility of nature" being a "common trope in the life sciences," throughout the 1960s genes were increasingly referred to (often interchangeably) as an alphabet, a language, a code, a message, and an inscription, a series of metaphors that were "rendered more poignant" in the popular imagination "in the context of the double stranded helix" (ibid.). This model of the genetic code assumed a universal legibility, as van Dijck explains: "applied to the structure of DNA, 'code' infers the

idea of a rule-governed system of communication that can be understood by everyone who has a key to its formative principles" (ibid.). The
concept of genetic code or language in the 1960s signaled a definite
shift "from metaphors of phenotype to metaphors of genotype," and its
widespread usage "far exceeded the function of exemplification proper"
(ibid.). In contrast to the 1960s, representations of genes from the 1990s
left behind images of the gene as factory and manager-controller from
early biotechnology, and moved to a set of associations which combined molecular biology with computer science. For van Dijck, "the
gene metamorphosed into the 'genome,' genetics into 'genomics' . . .
[and] more intricate mental concepts were needed to imagine the genome as a digital inscription of the body's genetic make-up" (ibid., 120).
This shift meant that rather than being seen as a unidirectional flow of
messages or codes, genes were increasingly conceptualized within more
complex, interactive models: "circulation, rather than a linear flow of
information, provided the vector for the dissemination of meaning . . .
the body became part of an informational network" (ibid., 121).

This reconfiguration in how the gene is conceptualized has wider
implications for how the body is imagined in general. Van Dijck sums
this up:

> The idea that the human body can be coded in a decipherable sequence
> of four letters, and hence in a finite collection of information, is based on
> the epistemological view that computer language—like molecular "lan-
> guage"—is an unambiguous representation of physical reality. Whereas
> the metaphor of mapping suggests an analog representation—a linear reg-
> istration of a flat surface—the sequencing of genes ushers us definitely into
> the digital era. Digital encoding differs from analog recording in imposing
> a language of zeroes and ones, combined into great complexities, onto
> the human material body . . . Through the inscription of "DNA-language"
> in digital data, the body is turned into a sequence of bits and bytes whose
> function is no longer exclusively representational. (Ibid., 123–24)

If "biotechnology combined with informatics transforms the body
as an (organic) object of knowledge into an ordered collection of biotic
components—itself an 'image' or 'concept,'" as van Dijck suggests (ibid.,
124), then the problem of reading the body presents itself in a new way.
If the geneticized body is understood as an informational network,
what new forms of insecurity enter the dialogic exchange of genetic

legibility? As Keller suggested over a decade ago, "the body of modern biology, like the DNA molecule . . . has become just another part of an informational network, now machine, now message, always ready for exchange, each for the other" (Keller 1995, 118)—but how does this re-configure our understandings of human agency in relation to biology?

Drawing on the work of Celeste Condit (1999), Patrick Gonder sug-gests that "human agency over the body" becomes an incredibly vexed question after Crick and Watson "broke the genetic code" in 1953 (Gonder 2003, 35). As Judith Roof has put it, the visualization of the DNA spiral as the "secret of life" produces a new image of knowledge in which by "stringing binary pairs of elements, DNA's material code works like a computer program—all complexity reduced to presence or absence, the ordering of switches, the discernible, identifiable, locat-able, enumerable operations of existence" (Roof 1996b, 173).[3] In this ap-parent opening up of humanity's reflection upon itself, "DNA's alternat-ing constituents suggested the possibility of a clear map of humanity's origin and mode of reproduction" (ibid., 172). Roof cautions us about the consequences of this shift in the discursive logic of coding, which extends debates about the meaning of the gene into the foundational structures of knowledge production.

The desire to make visible the hidden secrets of the body can be tracked through a long history of scientific endeavor, which culminates in the new promise of genetic vision. In her more recent work, Roof (2007) has gone on to argue that the notion of the gene as a code domi-nating popular conceptions of genetics is a structuralist metaphor with very particular consequences. Following Keller (2000), Roof argues that during the twentieth century, the gene came to be imagined as that which would supply the key to life's mysteries and became the perfect expression of the notion that to understand the world is to discern its meaningful structures and take them apart, or decode them. To think of the gene as a coded inscription makes the body a text for interpreta-tion, and thus conceived, the body is open to decoding and recoding: it becomes subject to authorship and regulation. DNA thus becomes "the agency through which we imagine that bodies are produced and changed" (Roof 2007, 24). According to Roof, this textualization of the biological as code brings with it a notion of genes as units of informa-tion that contain history and thus become agents of the future: "when we imagine genes as agents, they become literal representations of our bodies, our wills, and our desires" (ibid., 149). In this shift of emphasis from imagining genetics as a language to imagining it as a code, the geneticized body in contemporary culture moves into what Hayles (2005, 20) discussed as a "computational universe" that operates according to principles not of speech or writing, but of code, in which questions of interpretation take on a different significance from those at play in the signification of speech or in the iterability of the written text. With the shift from language to code, how is the concept of legibility itself transformed in the computational, rather than linguistic, universe of genomics?

The encoded body as a threat can be traced back to the "body rebellion films" of the 1950s and 1960s, as Gonder has argued: the body begins to be understood as "codified" in ways obscure to the nonscientist and is thus "construed as a site of suspicion, as a space for potential rebellion that requires special and extreme levels of surveillance by scientific experts" (Gonder 2003, 35–36). If throughout the history of the cinema, the woman's body has been the site of masculine desire and anxiety, in science-fiction films at this time it is the reproductive female body that is treated with particular suspicion. In this context, the gene is seen as regulating physical form through means that are "imperceptible and unpredictable" (ibid., 35). These films offer fantasies of controlling the unruly reproductive body through techno-scientific intervention into cellular activity.

In more recent films, the problem of coding translates into questions of bodily legibility, placing the geneticized subject in a series of investigative spaces inflected by technologies of surveillance to rehearse the new scenarios of intelligibility made possible by the informationalization of the body. Screening technologies designed to secure the truth about the body at a time of insecurity also generate fresh possibilities of fraudulent deception and impersonation: for instance, the iris scanning in *The 6th Day*, the fingerprint in *Gattaca*, the body's unique smell in *Alien: Resurrection*, and the personalized password in *Code 46*. Scientific expertise intended for the surveillance and control of the authenticity and singularity of human identity in each film is turned against the authorities and produces instead a proliferation of imitative techniques that generate further insecurity. The cinematic life of the gene works partly through the notion of the body as code produced through a diegetic sense of invisible cellular structures,[4] be it as the encrypting of the wisdom of an ancient civilization in *The Fifth Element* (Luc Besson, 1997); the strange, sexually transmittable computer virus in the protagonist's electronic DNA in *Teknolust*; or the threat to the traditional ban on incest posed by fetal cloning in *Code 46* (see chapters 6 and 8). The gene is given cinematic presence in the play upon the gaps between these screens: the desire for visual certainty and the fear of its impossibility, and the promise of seeing the truth about the body with its repeated elusiveness.

The blending of biological and cultural codes introduces problems of desirable and undesirable mixing through reproduction. In the body-rebellion films of the 1950s and 1960s, Gonder suggests that the mixing of biologies through genetic engineering is firmly embedded in cultural fears of racial mixing, of the "miscegenous body," and of the threat of the "potential to be nonwhite"; he argues that "the monster . . . that returns with such a vengeance is, at least on one level, the threat of racial difference or, more specifically, the problem and perhaps inability to determine racial difference" and that the "instability of the body itself" is at stake (ibid., 39). The horror in these films typically erupts from "the dissolution of corporeal homogeneity, the realization that the body is essentially polynucleated, a fragmentary collection of many potentially rebellious and independent units" (ibid., 33).

Whereas earlier films required the expulsion of the unhealthy body or the amputation of a dangerous body part in order to expunge the threat of racial indeterminacy, more recent films have produced the dynamic between geneticized and racialized bodies in rather different, though related, ways. Since the 1990s, the fixity and visibility of racialized bodies in film have articulated an opposition to the mutability of genetically engineered or cloned bodies, or what Paul Rabinow famously calls "biosociality": "a biologization of identity different from the older categories of the West (gender, age, race) in that it is understood as inherently manipulable and re-formable" (Rabinow 1999, 13). The tension between unchangeable and observable identities rooted in supposedly biological differences and the new, slippery potentialities of genetic recoding is epitomized in particular bodies representing different degrees of fixity, purity, or legibility: for example, the clichéd black-and-white masculinities of the two protagonists in *Evolution*; the "equal opportunities,"[5] one of each type of gender, race, and class in the final space-mission scene in *Gattaca*; and the team of scientists in *Alien: Resurrection*, in which the group represents the diversity of the human gene pool. In other films, visualizable racial differences suggest biological patterns to hidden genetic coding: for example, in *Casshern* (Kazuaki Kiriya, 2004) the biosocialities of genetic engineering are tied to the whiteness of the new race of engineered neohumans, and in *Code 46* cosmopolitan multiculturalism gives life to the problem of inappropriate genetic mixing. The blending of different cultures gives form to the artificial blending of different biologies in the fetal cloning techniques that generate the narrative problem of *Code 46*, in which the genetic code is made visible as the multicultural becomes the trope for desirable biodiversity.

In the noirish science-fiction film *Species*, the central problem is the invisible impurity hidden within the body of the female protagonist Sil (Michelle Williams and Natasha Henstridge as child and adult, respectively), who shifts between embodying the desirability of white perfection and displaying the monstrousness of mixed DNA. Putting this continuing concern with the regulation of biological purity through genetic coding together with the genre's more recent preoccupations with genomic surveillance and security techniques, the film makes the legibility of the codes of white femininity both highly desirable and ultimately elusive. If the enigmatic figure of the woman has conventionally driven the investigative impulses of narrative cinema, then the genomic encoding of her body extends the problem of solving her mystery into a "computational universe" in contemporary science-fiction films (Hayles 2005, 20).

The genetic dimensions of the tension between desirable purity and undesirable mixtures are played out through the juxtaposing of the fixed identities of the security team and the mutating monstrousness of the genetically engineered femme fatale whom they are employed to hunt down. The members of the team present an image of the diversity of the human gene pool, which is set in opposition to the mutating incarnations of Sil, the half-human, half-alien female.[6] In contrast to Sil's mutating surface illegibility, the team of experts assembled to join Xavier Fitch (Ben Kingsley) and capture her very visibly embody a rather clichéd, highly legible set of social codes: Preston "Press" Lennox (Michael Madsen) is the white, working-class tough guy; Laura Baker

10. The team of experts (in front) assembled to join Fitch and track down the monstrous femme fatale in *Species* (Roger Donaldson, 1995): Laura Baker (Marg Helgenberger), Preston "Press" Lennox (Michael Madsen), Steve Arden (Alfred Molina), Xavier Fitch (Ben Kingsley), and Dan Smithson (Forest Whitaker).

(Marg Helgenberger) is the white, middle-class female scientist; Steve Arden (Alfred Molina) is the naive professional Western intellectual (in this case, an anthropologist) with expert knowledge of other cultures; and Dan Smithson (Forest Whitaker) is the sensitive, black, middle-class empath (whose heightened feelings for others make him both a social outcast and a highly valued tracker). The repeated use of group shots of this ensembleof Fitch and his team (see figure 10) establishes the notion of genetic diversity through social variety as the team represents the range and strength of human qualities (physical, technological, intellectual, and emotional). Here, the surface transparency of social differences anchors cultural codes as stable and predictable, their very legibility contrasting with Sil's posthuman mutability, through which she evades capture. In this battle for superior powers of corporeal literacy, the old social categories (race, class, and gender) to which the human team clearly belong meet the new biosocialities of genetic engineering and cloning (Rabinow 1999) in the figure of Sil, whose threatening transgenic blend contrasts with the reassuring visibility of the team's social distinctions. The artifice of her hyperdesirable white femininity (any man, it seems, will say yes to her) disturbs the boundaries around the purity of the white body as it is revealed to hide the most unnatural of forces (see figures 1–3).

11. The many faces of the mutating, transgenic Sil (Natasha
Henstridge) in *Species* (Roger Donaldson, 1995): (1) hyperfeminine
white ideal; (2) half-human, half-monster; (3) predatory alien.

Species explores the converging techniques of artifice in the hyper-mediated urban cultures of consumption and in the computerized experiments of militarized science, posing the problem of the computational legibility of the geneticized body of the woman—or, rather, the fear of its potential illegibility. As with *Alien: Resurrection*, discussed in the previous chapter, the narrative problem posed here is one generated by the dangers of recombinant DNA. The genetically engineered female species of the film's title blends together the drives of nature and the compulsions of culture in one perfectly human, but utterly monstrous, body. Just as her senses are bombarded with the external imperatives of the contemporary media and consumer cultures around her, so her reproductive desires are driven from within by the genetic force behind her violent biological urges. Combining the seductive power of the femme fatale of film noir with the mutating reproductive incarnations of "the monstrous feminine" of science-fiction and horror films (Creed 1993), the transgenic protagonist simultaneously embodies the threats of deceptive female sexuality and of uncertain procreative becoming. Just as the film's tag line poses that double threat—"Men cannot resist her. Mankind may not survive her"—the much debated placement of woman as the cinema's enigma is given a biological form here through the notion of her embodiment of mixed genetic and generic codes.

If cloning and clothing seem interchangeable in the genetic imaginary of the young patient discussed at the beginning of this chapter, they become inextricable here in a rather different way. In *Species*, the survival of the transgenic species in an alien (human) culture depends upon biological and sartorial disguise: identity theft becomes an instinctive solution to the threat of extermination by the authorities who first engineered her. The narrative drive (to hunt down and destroy the threatening woman) is premised upon the potential for confusion, deception, and misinterpretation at the nexus of biological, linguistic, and cultural codes. Like numerous other science-fiction films—such as *The Andromeda Strain* (Robert Wise, 1971), *Alien* (Ridley Scott, 1979), and *Contact* (Robert Zemeckis, 1997)—*Species* uses an unexpected extraterrestrial communication to generate the events which lead to the exploration of new frontiers of science. In this case, an unfamiliar DNA sequence sent by an unknown source from outer space leads to its experimental combination with human DNA, producing a female whom her scientist progenitors call Sil. The genetic code—masquerading as a

potentially unifying, if not universal, language—masks a hidden malev-
olent intention from an alien intruder that builds miscommunication
and misrecognition into the diegesis from the start.

The desirable legibility of the genetic code generates a rationale for a
proliferation of screening technologies throughout the film. As in *Gattaca* and *Code 46* (see chapters 5 and 6), the mise-en-scène in *Species*
is organized around a series of scientific and surveillance techniques
motivated by the desire for transparency through visual legibility. The
power of techno-science in *Species* is tested against the ingenuity of the
threat to humanity which science itself has produced and must now
destroy. The narrative of the chase is punctuated by laboratory scenes
displaying new genomic procedures. As in *Jurassic Park*, a pivotal early
scene in *Species* shows the head scientist (Fitch) explaining genetic-
engineering procedures to the diegetic audience of his assembled team,
accompanied by visualizations of the procedures he describes on the
screen.

The narrative's trajectory is organized around the tension between
the desire for science to make visible what it seeks to experiment upon
and control, and Sil's apparently instinctive ability for disguise and camouflage. A multitude of screening devices (visual, informational, and instinctual) are mobilized by each side to enhance its visual-literacy skills
(see Smelik 2008) and to make the other's codes more transparent and
predictable. The film generates (and repeatedly thwarts) the desire for
the genetic code to provide the key to human reproductive powers and
thus to the mystery of life itself. Usurping the procreative capacity of
the female body, genetic scientists engineer a new life-form by combining the unknown genetic code (from outer space) with human DNA.
The outcome, Sil, looks like an innocent child but hides a deeper, monstrous, biological truth beneath. Just as the figure of the woman has
been the traditional site of deception and suspicion in the cinema, so
her physical interior has been imagined as the space of conquest and
discovery in science (Shohat 1998). The rendering of the female body as
code promises the chance for definitive knowledge and control.

The regulatory desire that seeks to control the body through vision
is built into the architecture of the laboratories, promising but failing
to deliver genetic transparency throughout the film. In the opening
scene, a vast, high-tech laboratory displays multiple sources of visual
control in the panoptic architectures of observational regulation that

12. The child Sil's glass cell in Fitch's panoptic laboratory
in *Species* (Roger Donaldson, 1995).

give the scientist a godlike vantage point from which to look down on
his apparently vulnerable subject (the child Sil), whose glass-walled cell
affords him full visual access (see figure 12). The visual image of the girl
inside the glass prison who is being continuously recorded is displayed
on the computer screens of the observation platform above. This mise-
en-scène of totalizing control through maximum technological vision
gives architectural shape to the desire for the genetic code to make the
human body transparent and manipulable. As Sil's extermination by
cyanide gas is executed with military precision and technical accuracy,
the panicked (but inaudible) reaction of the girl is first observed directly
through the glass walls by Fitch, the scientist who engineered (and fathered) her. The sequence of shot and reverse-shot looks establishes
an intimacy between them, and, unable to witness her death with his
own eyes, Fitch turns to watch the outcome of his final solution mediated by the computer monitors. Once hidden by the cloudy gas and out
of the control of his direct vision, however, the elaborate architectures
of surveillance are suddenly matched by the physical force of the girl's
body, when the young victim turns bionic escapee. In a spectacular disruption of this relay of controls and military security, the child dives
through her cell wall and runs to freedom outside the base. Thus, with
all their panoptic brilliance, the visualizing technologies of science cannot contain her genetically engineered body and make it legible, predictable, and regulatable. Such a fantasy is thwarted by the speed and

force of the transgenic species that science itself has engineered, and of which it has lost control.

The narrative tension that follows in the film is built around the battle between the combined expertise (military tracking devices, digital visualizing techniques, and further genetic engineering alongside human instinct, intuition, and brute force) of the team hired to capture the rapidly maturing species before she mates, and her swiftly expanding databank of resources for disguise and deception to avoid capture (images from the Internet, television news, and advertising; sartorial styles of mannequins in shop windows and of other women in the street; and her bionic physical power, mutational body, and alien instinct for genetic matching to maximize her preservation). Questions of automated femininity as a project of decoding and recoding are placed center stage in the power struggle between the intellectual, emotional, and physical force of the genetically human team and Sil's instinctive ingenuity and camouflage techniques. The driving narrative question becomes: can they decode her before she can recode herself? Her genetically engineered body is thus the focus of an epistemological crisis about whether seeing really is believing, and about what counts as evidence in the quest to secure bodies as biologically legible. The status of visual evidence and of the relationship of vision to knowledge is rehearsed in different ways through the phantasmatic place of the woman's body in the histories of both the thriller and science-fiction film genres (and indeed in the cinema's preoccupation with its own status as a medium); here these questions become genomic ones, as the geneticized body provides the imaginary space for a revitalized desire to decipher femininity's hidden truths through DNA. The cinematic life of the gene in *Species* is produced through an interrogation of the incommensurability of technological vision, knowledge, and the biological truth of the woman's body, as the film reiterates a familiar concern about its own signifying credibility through its preoccupation with the image cultures of urban consumerism. The architectures of surveillance that govern the opening of the film match two later scenes. The first displays the perfect match between the visualizable and the genetically legible, though only to accompany a story of the failures of each to contain the product of their union. In the scene in which the team of experts employed to hunt Sil are given their briefing, Fitch recounts the story of her origin through a series of magnified images: the double helix of DNA with its familiar lettered pairings; the microinjection of the human ova with the alien DNA; and the rapid cell growth of one successfully engineered egg, as we watch the familiar divisions of the miracle of the beginning of life (one cell becomes two, two become four, and so on). These familiar visualizations of science are followed by images of the orphan child filmed awake and asleep (scientific closed-circuit television footage) selected to accompany Fitch's account of the indicators of alien biological disturbance which continue to necessitate her termination. The swift transitions echo the alarmingly fast biological process they attempt to document: the embryo is formed in hours and the fetus in days, and the four-month-old baby resembles a six-year-old child (we, like the diegetic audience, hear Fitch's narration and watch the miracle of genetically engineered life demonstrated in a matter of seconds). Here the biological and the cultural accelerate each other, as these images of the speed of physical transformation by this unnatural biological force expose its transgenetic abnormality through the rapidly edited visual sequence on the screen.

The display of genomic expertise here, however, also reveals the problematic status of this visual evidence of life itself, blending narrative authority with scientific imaging techniques to tell a story of their failure. Fitch's illustrated narration to his assembled audience of experts in the laboratory invites the cinema audience to reinterpret the cyanide extermination in the opening scene discussed above. The apparently horrific immorality of these events (part Holocaust replay, part child abuse) is reframed as they are rationalized retrospectively as reasonable and justifiable preemptive actions to save humanity, and maybe even the child, from the threatening alien life lurking within her body.[7] But the brutalities behind these risky genetic experiments and the forms of surveillance they require linger beyond Fitch's neatly executed rationale, which transforms him from ruthless father to responsible scientist, and Sil from defenseless child to potential biological threat. As the film progresses, Fitch's authoritative narrative of monstrous DNA versus innocent science is undermined by the emergence of obvious links between Sil and her hunters through a series of tropes. Each member of the expert team is linked to the monster they hunt in a particular way: Fitch's inhumanity toward his offspring, Sil (the tears at the prospect of her death are his only emotional moment in the film) and later toward members of his own team (Press and Laura, whom he refuses to save because of protocol) match Sil's ruthlessness; Press shares her capacity for brutality and her instinctive drive for self-preservation; Laura's appearance is twinned with Sil's classic white, blonde femininity, and they share a heterosexual desire for Press; Steve is ultimately her genetic match and successful reproductive victim (the father of her offspring); and Dan's intuitive empathy with Sil's suffering brings the team in ever closer physical proximity to her, although his naturalized emotionality contrasts with her lack of it. As in *Alien: Resurrection* and many other science-fiction films about various experiments with artificial life, emotions are the marker of the human in *Species*, proving scientists to be as horrific as the monsters they produce. In both *Alien: Resurrection* and *Species*, lack of empathy places the human status of the white female protagonist in question: but while humanness is restored to Ripley through her traumatic encounter with the horrors of her origin, Sil's lack of emotional register continues to mark her inauthenticity: conventional feminine intuition is located instead in the black male empath, which stereotypically connects him to nature through feelings as he exceeds the usual capacities of human sensitivity and emotional literacy.

A second scene in the film that echoes the earlier scientific crisis of the relationship between vision, knowledge, and bodily legibility is the one in which the team attempts to grow pure alien DNA, motivated by a desire to see beyond the mask of Sil's genetic combination with human DNA. The scene cuts between the live microinjection in the laboratory and the adult Sil, channel hopping in a motel room in Los Angeles. The close-up image of microinjection (the prosthetic penetration of the human cell by the pipette through the directing hands of the scientist) has, as Sarah Franklin (2002) has argued, become an iconic emblem for the life-generating capacities of genetic engineering. In this scene, the abrupt cut from close-ups of pornography and an advertisement for hair dye on the television to the shots of the computer screen showing microinjection places images of biological and cultural artifice in almost clichéd dialogue. Articulated through a mise-en-scène of techno-scientific apparatuses at their most digitally powerful, the desire to master life in this scene is enacted on computer and cinema screens simultaneously. A highly mediated view of the microinjection conducted by remote control is magnified on one screen through the other, as the female scientist narrates the story of the new life-form she is engineer-

13. Laura Baker (Marg Helgenberger) engineers a new life-form as the other members of the team of experts—Dan Smithson (Forest Whitaker), Preston "Press" Lennox (Michael Madsen), Xavier Fitch (Ben Kingsley), and Steve Arden (Alfred Molina)—view a microinjection magnified on a computer screen in *Species* (Roger Donaldson, 1995).

ing and the other members of the team (all male) turn expectantly to witness what they, and the cinema audience, assume they will see with their own eyes: genomics in action (see figure 13).

But, as in the opening scene, on the threshold of scientific certainty, as image and truth are about to unite, technology fails: this time, the camera goes down, and both computer and cinema screens go blank. The diegetic and extra-diegetic spectators share a loss of point of view in the moment of technological failure. Neither audience can see the state of the microinjection procedure, and when the computer's camera is switched back on, the two groups are simultaneously shocked by the monstrous outcome of the successful penetration of the cell by the alien DNA—a rapidly proliferating, amorphous, brown cellular mass—made all the more shocking because the image of the moment of cell penetration has eluded them. The slowing down of time in the moment of suspense around the microinjection contrasts with the speeding up of the pace when the rapid division of the alien cells far exceeds the pace of normal biological development. As the creature grows before our eyes, it begins evolving into different forms: from a frothy cellular mass through a sticky, pulsating malignant growth to a reptilian beast with tentacles. The horror of this bubbling, cancerous mutation underscores our sense of disbelief: how could the procedure have been completed without our knowledge and beyond our vision? Echoing the earlier mismatch between the scientist's intentions (the generation of life through genetic engineering) and the possible outcomes (a terrifying and unstoppable life-form), the failure of the screening techniques here makes visible the crisis in the teleology of vision, knowledge, and legible bodies despite desires for genomics to guarantee the opposite.

This interrupted teleology of engineering life in the scientific and military laboratories works in contrast to the capacities of the artificially engineered Sil. Unlike the expert team with its failed attempts to generate an imitation of alien life through cellular manipulation in the laboratory, Sil successfully imitates the images of life she encounters through multiple screens in the film: a stolen portable computer on the train, shop windows in the streets of Los Angeles, and her television in the motel bedroom.[8] Whereas the team proves unable to use the image on the screen to control the process of engineering life, she successfully incorporates everything she sees on the screen into her strategies of self-preservation. Sil's imitative biological origins translate into her ability to copy screened images: the hair dye from an advertisement, the sexual seduction from pornography, and the car crash from the news. All these images translate into information for her escape and reappear as resources for her own engineering of events later in the diegesis: the car crash is both real (the car explodes, and a woman dies) and a deception (Sil plants a body part in the car). Like a biological scanner, her ability to read visual images as data and render them live is matched by the way in which she reads live bodies as informational codes. She uses this mimetic capacity in her secret observations of the expert team. She is invisible through the camouflage of her biological and cultural disguises, and point-of-view shots show the human detectives looking for her as she watches them like a voyeur or a peeping tom. Mimicking their actions and sometimes even their speech (she almost seems to ventriloquize Fitch's words), she imitates their plans but reverses their desired outcomes in her copycat schemes to escape. As she sets false trails and leaves her emotional scent for the empath Dan to follow, her computational accuracy of prediction and control of data processing far outstrips the technological skills of the scientists and the screening techniques of surveillance. Importantly, Sil herself cannot be captured as an image: both the closed-circuit television at the motel and the

14. From train conductor to bride—the sartorial disguises of Sil
(Natasha Henstridge) in *Species* (Roger Donaldson, 1995).

home-video camera of one of her victims fail to produce a clear likeness of her. Genomics thus fails to provide the techniques to control life by producing it as a predictable and observable biological code, inventing instead a life-form whose capacity to translate the world into information far outstrips the ambitions of the scientist's original desires.

The conventional place of woman as image in contemporary consumer culture is underscored by the casting of a model like Natasha Henstridge as Sil, but it is also scrambled by Sil's modes of visual literacy that accompany her avoidance of being captured on camera. As random as the stolen food on which she binges before her monstrous metamorphosis into an adult body, these visual images bombard her (and us) with a dizzying array of arbitrary imperatives about how to do femininity. Like an embodiment of the posthuman disregard for material specificity, Sil ingests these images as if they were organic matter (see Hayles 1999). Her compulsive appetite for both food and images is driven by a biological imperative.[9] This convergence is emphasized by the simultaneity of Sil's accelerated biological and cultural mutations, which facilitate her disguise.[10] Again, cloning and clothing coincide as Sil improvises feminine desire through commodity consumption: she becomes a mistress of the masquerade, stealing and borrowing whatever cultural accouterments she needs (money, credit cards, cars, names, and clothes), as she appears as train conductor, bride, and sexual predator (see figure 14). According to her reading of the cultural codes around her on the streets of Los Angeles, femininity seems strategic and instrumental; it can be incorporated through the popular culture that provides an endless database for random reassemblage. The ubiquity of images of commodified femininity makes impersonation endlessly possible for a novice at the task. Since femininity itself seems to be *only* a set of conventions, easily performed in a single gesture (take your top off in a bar, ask a guy to take you home, and the answer is predictable), imitating the human female proves immediately possible for Sil as hybrid stranger with a radar-like ability for assimilation and adaptation.

But it is Sil's computational universe, in which she reductively reads everything as data, that ultimately is her vulnerability, exposing her outsider status and leaving behind clues of her crude approximations for the team of experts to follow. Ultimately, she reads the artifice of femininity too literally, as if it were purely information and not also a dialogic identity. This inability to imitate femininity appropriately and produce it contextually (such as her wearing a wedding dress to go shopping) exposes the limits of Sil's mode of reading culture as informational: although femininity can be performed through the conventionalized repetition of gesture (Butler 1990), it nevertheless has to become an ontology in order to become fully social. Reading the femininities of consumer culture as *only* surface images, providing informational codes and not embedded as cultural practice, Sil's cultural illiteracy begins to contrast with her biological literacy, which she deploys with unfailing intuitive accuracy in her pursuit of a genetically suitable mate. In her quest to find a mate, Sil reads men's DNA simply by sensing their pedigrees through physical proximity. Her urgent sexual seductions are driven by reproductive necessity. Propelled by this instinctual drive for the survival of the fittest and an accelerated biological clock, she rejects imperfect males before intercourse, using her alien capacity for violence to avoid the usual reprisals for such a transgression of expected feminine acquiescence. As the embodiment of the conventionalized ideal of white beauty (blonde, blue-eyed, slim, tall, and young) she is guaranteed any heterosexual mate she chooses, but she performs the femininities necessary for her reproductive quest utterly strategically and with a ruthless, masculine detachment. In a reversal of the usual gendered seduction scenarios, Sil's predatory sexuality, motivated by her sense of satisfactory biological data, turns the male body into a code which is simply there to be read and used for her reproductive convenience.

Here, anxieties about male redundancy and dispensability in a future of powerful, autogenerative females combine in the double threat of the femme fatale and of the monstrous feminine.

In *Species* genetic engineering and cloning techniques become forms of biological design whose artificiality is articulated through the cultural disguise of impersonation and produces problems of illegibility. Just as Sil ultimately fails to impersonate humanness through femininity, so the scientists in the film are thwarted in their attempts to imitate biology, to control life through genetic engineering. Sil's literal readings betray the truth of her genetic artificiality; the experts' inability to see the moment of microinjection of the alien DNA betrays the fallibility of both the humans and their computational technologies to see what is really going on at the genetic level. In the moment of technological failure, their desire (and perhaps ours) lingers, exposed. The array of human and alien misreadings throughout the film shows each side's failure to shift appropriately between the different registers of legibility required by codes and modes of culture, biology, and techno-science. The scientists misread transparency as certainty (for example, the transparency of Sil's cell walls does not guarantee security), revealing their misplaced belief that knowledge and vision guarantee each other; Sil misreads culture as pure information (images, bodies, and clothes are only resources), suggesting that her genetic makeup translates into a computational subjectivity that perceives the world as bytes of data.

Like a number of other "species-level biohorror" films (M. Hills 2005),[11] *Species* delivers a monstrous femininity that embodies fears about the future reproductive status of the masculine subject in the posthuman culture in which "information lost its body" (Hayles 1999, 2). As Judith Roof puts it, what does "paternity, parentage, lineage, and generation consist of" if genetic science now offers "a more reliable mode of identification than blood typing's phenotypical groups" (Roof 1996a, 172)? For Roof, this change from the symbolic power of the name of the father in language (the Lacanian Law of the Father) to his literal DNA print that has come to confirm his claims to paternity represents a transformation in how the paternal signifier can exercise ultimate authority through its metaphorical power: "If the Name-of-the-Father stood for lineage, DNA technology threatened to define the father as an enumerable series of chained signifiers that account for every fact,

aspect, clinical operation, and anatomical morsel of human existence."
In this shift from symbolic metaphor to genetic metonym, "DNA, it
seems, provides the literal end to the problem of symbolic crisis" (ibid.).
The significance of this change in how paternal authority is conceptualized is profound for Roof: insofar as "the Name is replaced by the
comprehensive code; the nothing secured by the name's metaphor is
supplanted to the too-much something of an overextended set of signifiers which replace metaphor with fact and Law with code" (ibid., 173).
Such a shift reconfigures the place of paternity in its traditional biological and cultural forms. Roof's reading of the aspirational certainties of
the genomic shift from law to code (or from metaphor to fact) presents
an important challenge to theories of how paternity works to secure
patriarchal authority. But in *Species* the fear is focused on the notion
that even this biological code might prove elusive or illegible. The repeated failure of humans and their technologies to read Sil's codes correctly and to control her through making her biology visible puts the
certainty of the code, or at least its facticity, into question. In short, the
film presents the monstrous implications of the genetic code's not being intelligible (as stable and transparent fact) for the future gendering
of embodied reproduction.

If genetically engineered femininity cannot be decoded, its dangerous indecipherability threatens the potency of its progenitors. Here,
the monstrous figure of Sil moves beyond the code as literal fact and
into the domain of the slipperiness of interpretation. If genomics has
produced bodies requiring more sophisticated modes of reading than
their predecessors, how can science sustain its authority and expertise?
This is the risk of the mutational bodies of transgenic cloning that pose
new threats to the powers of science to control their legibility through
vision. Sil's capacity to escape the scientists' visual literacy suggests a
genomic embodiment more akin to literary criticism than to language.
We might follow Robert Pollack, who suggests that we take literary criticism as a model for understanding genetics in such a way as to push
reading beyond the model of deciphering: "the cells of our bodies do
extract a multiplicity of meaning from the DNA text inside them, and
we have begun to read a cell's DNA in ways even more subtle than a cell
can do" (quoted in van Dijck 1998, 154). As van Dijck argues, the model
of genetics as literature rather than language or code opens genetics
up to questions of multiple interpretations and contested signification:

"the inherent polysemy of DNA texture enables multiple interpretations of the same letters" (ibid.). Pollack pushes his challenge to the literal model of reading still further when he argues that deciphering the "meaning of life" is never simply a translation but rather the production of a "consensus trans*literation*" (quoted in ibid., 155). In repeatedly thwarting any stable ground for such consensus, Sil's mutational femininity gives the lie to any such illusion.

Within this framework, cracking the code to decipher the Rosetta Stone (the artifact used by Champollion in 1821 to interpret Egyptian hieroglyphics) is not an appropriate metaphor since, as van Dijck points out, "the Rosetta Stone was a rebus, a hieroglyphic sculpture that carried different language inscriptions of the same text" (ibid., 155). In this sense, femininity as code in *Species* might be recast through the notion of the hieroglyph. As Mary Ann Doane has argued in a very different context, hieroglyphics might be thought of as representing both the most obscure and the most transparent of communication systems:

> On the one hand, the hieroglyphic is summoned . . . to connote an indecipherable language, a signifying system which denies its own function by failing to signify anything to the uninitiated, to those who do not hold the key . . . On the other hand, the hieroglyph is the most readable of languages. Its immediacy, its accessibility are functions of its status as *pictorial* language, a writing in images. (Doane 1991a, 18, emphasis in original)

As a monstrous femme fatale, Sil's duplicity in *Species* is given a hieroglyphic potency because, like the gene itself, she represents both the most illegible of ciphers and the most readable of signs. Sil's legibility is both self-evident and beyond all technological capability, placing anxieties about the biological transparency and authenticity of the body alongside the cinema's concern with its own visualizing and fictionalizing powers. The cinema promises transparency through its techniques of verisimilitude and the audience's proximity to the image; yet it also disguises the imitative purpose of its generic repetitions. The problem of whether seeing is believing in this film is played both ways within visualizing technologies that depend upon the fetishistic disavowal of "I know, but . . ."[12] The cinema generally produces a desire for transparency alongside the impossibility of its delivery with a certain pleasurable repetition: its own history is one of depth affect through surface image.

The figure in *Species* of hyperdesirable white femininity as dreaded ruthlessness, sexually and biologically exploiting men, rewrites the duplicitous femme fatale (whose traditional desirability hides murderous intent) as the indecipherable posthuman cyborg, whose robotic, gendered perfection masks a terrifyingly selfish reproductive drive. The film might thus be read in the context of wider anxieties about the redundancy of the white male in the age of not only reproductive and genetic technologies but also of feminism. The dread of white, heterosexual males that their ultimate fate is to be only a host body for healthy sperm (as demonstrated by Sil's serial sexual instrumentality) gives this struggle for control through visual literacy a racialized and sexualized significance that underscores anxieties about the changing cultural codes of reproductive heterosexuality. Whereas the classic femme fatale of the 1940s has been read as condensing wider fears about changing forms for families in postwar culture (Harvey 1998), the monstrously fatal feminine clone in the 1990s, passing as sexual ideal, embodies the terror of posthuman reproductive potentialities. In a film in which a genetically engineered monster consumes images like food and scans men's bodies with computational accuracy to assess their procreative pedigrees, the disturbing artificiality of both the cultural and biological codes are epitomized in the reproductive embodiment of automated femininity.

The story of linguistic and conceptual substitution (of clothing for cloning) with which I began this chapter enacts the problem that lies at the heart of the disturbances which structure the genetic imaginary: the desire for transparency produces a proliferation of confusion. In displacing genetic sameness onto sartorial sameness, the girl in the story reads biological depth as cultural surface. Duplication of genetic data (cloning) is imagined through duplication of appearance (uniform) in such a way as to make transparent the absence of difference that might mark out some individuals (like her sister) as special. Sameness spells safety if it can be made visible. If kept in sight, the transparency of sameness can be guaranteed. In this fantasy of "doing" clothing when she grows up, the girl reveals her wish to control difference (like the headmistress makes people wear the same clothes) and to have "everything" (being in two places at once) by cloning herself. The wish to abolish the threat of difference enables her to imagine her own desire fulfilled. If heterosexual reproduction can be bypassed (you don't need a mother and father) in favor of self-replication through the new paternal authority of science (you just need a scientist, a man) in a place where individuality is no longer a threat (no one is special), then the girl can fulfill her desire to have everything: she can be in charge of her own destiny and still blend into the crowd. While her individual agency is guaranteed through reproductive authorship (like a male scientist, she will make all these babies), the threatening specialness of others (such as her sister's capacity to entrance her mother) is disavowed through the new potentialities of replication in genetic engineering. This imagined shift moves her from the powerless child (subject to the authorities of school and family) to controlling adult (she will regulate reproduction, uniforms, and parental affection). In her fantasy, sartorial uniformity, like its genetic counterpart, will provide the justice of transparency through the new, visible laws of sameness.

But this young girl is not alone in her confusion. While her story may reveal her own particular unconscious investments in overriding traditional family structures, the substitution of clothing for cloning has wider imaginary purchase, as this chapter has shown. The displacement of the biological onto the cultural makes transparent the hidden threat of sameness within scientific genetic manipulation. As in the girl's fantasy of cloning in Phillips's account, in *Species* the genetically engineered girl has no biological parents but instead is produced by a male scientist, who narrates the story of her miraculous conception. And, again reminiscent of Phillips's story, Sil's genetically engineered duplication is expressed through her clothes: her transgenic mutations are echoed through her endless changes of identity using sartorial disguise. Sil's ideal female body is the register for confusions around legibility: the disguises at her disposal enable her to camouflage the monstrous feminine through the duplicity of the femme fatale. What we might call her mutational femininity articulates her genetically engineered body as the site of deception, warning of the dangers of interference at a deeper level.

Genetic engineering and cloning pose a threat to the transparency of identity and to its legibility by experts, sexual partners, and colleagues. *Species* investigates the potentialities and limits of visualizing technologies and information technologies to make legible the genetic truths of identity. The power of human ingenuity and the tools of cultural design are pitted against the power of genetic engineering and biological design to transform the fundamental meaning of life itself: the monstrous femme fatale develops techniques of cultural disguise paralleling her biological conception, which put her one step ahead of the experts who designed her. The potential for deception is placed at the heart of the narrative that juxtaposes the power of scientists to guarantee authenticity with the endless potentiality for interference, imitation, and duplication. The moments of apparent visual certainty on the screen (the microinjection of alien DNA to reveal Sil's disguise) rehearse what we might call the scientists' desire for genetic transparency: the power to visualize biology as pure information. However, transparency through visualization repeatedly eludes them.

If the confusion of cloning with clothing symbolizes the desire to make visible on the surface of the body the interference at the hidden genetic level, then the status of the image is crucial to the structuring anxiety at stake (and to the reassurance sought). The cinematic language of the gene searches for substitutions while simultaneously cautioning against the certainty the image promises. *Species* questions the power of visualizing technologies to secure genetic certainty, since genetic information cannot be made transparent by screening techniques (in all senses of the term). In requiring others to complete the dialogic circuit of identity production, the problems of competing interpretations and varying registers of legibility become apparent. The desire to transform the body into legible information (to contain the threat of sameness by making it visible), which can be read as pure data and replicated accordingly, is thwarted by the unexpected uses of technologies of imitation. Here, anxieties about the legibility of corporeal surfaces and depths are displayed across a matrix of informational and biological registers, and the transparency of the screened body remains a dubious and unfulfilled wish.

〔 **2** 〕

Imitations of Life

Nature creates similarities. One need only think of mimicry. The highest capacity for producing similarities, however, is man's. His gift of seeing resemblances is nothing other than a rudiment of the powerful compulsion in former times to become and behave like something else. BENJAMIN, *Reflections*

[4] Cloning as Biomimicry

As yet we have no adequate language through which to conceptualize the prospect of cloning. It represents something unassimilable. We cannot grasp the idea of a clone of ourselves any more than we can psychically fully confront the prospect of our own mortality. And if we try to imagine others (whether loved or repudiated) as cloned figures, the sense of disassociation multiplies. To some extent, the outcomes of genetic engineering find an obvious home in the vocabularies of imitation, since the figure of the clone is the youngest in a long genealogy of doppelgängers who have populated our imaginative landscapes and haunted our psyches. Like the identical twin, the clone embodies an uncanny synthesis of sameness and difference, defying the conventional demarcations around the singularity of the human body. But the clone goes beyond the limits of previous modes of conceptualizing imitation, presenting as it does a new techno-scientific intervention which returns us to cultural processes through its imitation of biological ones. Unlike the identical twin, the clone is a product of intentional interference with biological processes, an interruption to so-called natural generative patterns.

In the contours of the genetic imaginary in this book, two dynamics are present: the clone is configured through a familiar repertoire of imitative tropes, and its uncanny prospect disturbs us in new and ungraspable ways. When we try to imagine a future of clones, it feels beyond belief. The clone is a figure that still belongs to the world of science fiction, yet it takes center stage in the genetic imaginary through the announcement of its rapidly approaching physical presence in the laboratories of the key players in the global race of genetic experimentation. To use the word "clone" here is to concretize a fantasy figure that in its human form has yet to materialize. Yet the distinction between science fiction and scientific practice is a blurred one in the cultural imaginary, and the clone as a figure hovers on the horizon, combining the unconscious fears and desires that structure the phantasmatic worlds of both science and the cinema.

So how might we think about the clone in the cultural imaginary, through conceptual histories of imitative figures? Western science has long yearned to control nature by copying its laws, but genetic engineering and cloning place fresh demands on the metaphors we have relied upon to theorize such a relation. As the doubling of an organism, the clone might be said to embody and epitomize the imitative desires informing the genetic project more generally. Given that genetic engineering has enabled science to imitate nature with a new intensity, how might the existing vocabulary of mimesis, mimicry, masquerade, imposture, impersonation, and so forth provide a language for conceptualizing such usurpation? The figure of the clone represents a genetic double whose biological duplication must make its sameness visible to confirm its presence. This new form of replication defies the relations of self and other upon which the subject has depended for its sense of singularity. This chapter considers in what sense the clone is mimetic, and in what ways the possibility of its production stretches such long-standing notions beyond old limits. Throughout, I assume the clone to be a figurative phenomenon. I consider how to conceptualize the ways in which the clone undermines the conventional teleology of original and copy. If we are now construed as mere imprints of our genealogies (copies of our parents' combined genetic inheritance), does cloning produce the latest manifestation of the notion of the original as always already a copy (Butler 2002; see also Butler 1990)?

The word "mimesis" refers to imitation in art, literature, and biology. Genetic engineering does not introduce the biological into debates about cultural mimesis; rather, it returns us to the ways in which biological processes have informed how imitation has been conceptualized as a cultural process. As Elizabeth Grosz argues, Lacan's famous theory of the mirror stage (as well as his theories of the imaginary and of psychosis) is influenced by a model of imitation elaborated in Roger Caillois's analyses of mimicry and camouflage in the so-called natural world, which Lacan uses as the basis for his spatial model of a particular form of human psychosis (Grosz 1994, 46). In a 1935 essay, "Mimicry and Legendary Psychasthenia," Caillois (2003) rejects the functionalist approach to mimicry as a form of instinctive self-preservation in animals that casts camouflage as an adaptive response to their environment in the face of predators. Instead he suggests that an insect's capacity for morphological imitation might be analogous to the psychosis described by Pierre Janet as "legendary psychasthenia," in which space is perceived to devour schizophrenic subjects:

> Space chases, entraps, and digests them in a huge process of phagocytosis. Then it ultimately takes their place. The body and mind thereupon become dissociated; the subject crosses the boundary of his own skin and stands outside of his senses. He tries to see *himself, from* some point in space. He feels he is turning into space himself—*dark space into which things cannot be put.* (Caillois 2003, 100, emphasis in the original)

This *"depersonalization through assimilation into space"* blurs the boundaries between the organism and its milieu (ibid., emphasis in the original). Rather than following the drive of self-preservation typically attributed to the processes of camouflage, here life seems to lose ground—or, as Caillois puts it: *"life withdraws to a lesser state"* (ibid., 101, emphasis in the original). By offering a different model of camouflage in the animal world, one whereby the insect surrenders itself to the spatial demands of its environment and renounces life rather than simply adapting to it, Caillois produces a new way of thinking about the loss of self in camouflage and psychosis. Both, he suggests, produce a mimetic morphology, involving the occupation of space from outside of the self, and from outside (or from the other side of) the senses.

The permeability of the ego here is indicated by the resemblance of form between the organism and its environment. Like the ego of the

psychotic, the insect which camouflages itself and blends with its surroundings loses a sense of itself as the "origin of the coordinate system"
and becomes "one point among many. Dispossessed of its privilege,
it quite literally *no longer knows what to do with itself*" (ibid., 99, emphasis in the original). As Grosz suggests, this conceptualization of mimicry places spatiality at the heart of subjectivity (or its absence) and is
key to Lacan's theory of the infant's earliest sense of its own imaginary
otherness in the mirror stage (Grosz 1994, 46–50). As she argues, for
Lacan, "the biological body, if it exists at all, exists for the subject only
through the mediation of an image or series of (social/cultural) images
of the body and its capacity for movement and action" (ibid., 41). Thus,
in the first instance, if the infant is to be able to "conceive of itself as an
object, and a body, and if it is to take on voluntary action in conceiving
of itself as a subject," then it must develop "an image or *Doppelgänger* of
the body" (ibid., 41–42). Generally, Lacan is believed to suggest that in
this early stage of development, the infant recognizes itself as separate
and distinct from its mother only in the image mirrored back to it. This
image of its totality, or the unifying image of the body as a whole and
coherent, forms the basis of an imaginary anatomy, providing an "anticipatory ideal of unity" to which the "ego will always aspire" ("preserved
after the Oedipus complex as the ego ideal"), but which is at odds with
"the turbulence and chaos occasioned by its motor and sensory imma-
turity" (Grosz, ibid.). For Lacan, this imaginary identification is "fraught
with tensions and contradictions insofar as the child identifies with an
image that both is and is not itself" (Grosz, ibid., 42).

Seen through a psychoanalytic lens, the clone figure represents the
fantasy of a sense of oneself from the outside in one's entirety, perhaps concretizing the lingering imaginary image of idealized totality
described by Lacan. As Carole-Ann Tyler puts it so succinctly, since
we cannot "see ourselves seeing ourselves . . . the Other is the support
for the subject's self-image; in Lacan's phrase 'I is another'" (C. Tyler
2003, 194). Following this logic for a moment, perhaps the haunting
fascination of cloning for us is that it represents a bodying forth, or a
biological instantiation, of the self-other relation of the imaginary mirror stage: that which is always lost to us, but upon which our illusory
sense of a unified self is founded. In externalizing that relation and giving it a physical presence in the world, the fantasy of cloning promises
us an otherwise impossible perspective: a view of ourselves from the

outside, while we simultaneously continue to inhabit the body from within. Doubling our presence in the world, building on our sense of the uncanny identical twin, the clone violates our perceptual sense of imaginary interiority, robs us of a bounded sense of singular embodiment, and perhaps returns us to the compelling, yet disturbing, delusions of an early formative stage.

Another way to think about this is that the clone represents the external spatialization of psychic embodiment: in an appealing yet disorienting way, cloning biospatializes the self-other relation. The clone may not be entrapped and digested by space, as in the mimicry of legendary psychasthenia, but he similarly "crosses the boundary of his own skin" and "stands outside of his senses," losing the origin of his coordinates and becoming instead one point of a twinned dyad or of potentially endless duplications; seeing himself "from some point in space," the clone is dispossessed of his singular and privileged perspective and literally no longer knows where to put himself (Caillois 2003, 100–101). The clone faces us as an externalized embodiment of a psychic division, a doubling, which transforms our perception of our own interiority. In contemplating cloning, we are presented with a morphology which previously remained an imaginary imago. It thus not only threatens our investment in our singularity and particularity, but it also endows a concrete completeness to our idealized totality which is both desirable and unthinkable. The clone represents not the pleasing ideal of imaginary unity (of the mirror stage), to which the subject endlessly returns after its entry in the symbolic realm through its identifications with idealized egos, but rather the literalization of the contradiction at the heart of this love affair—or even the disorientating *"depersonalization through assimilation into space"* of psychosis (ibid., 100). Confronted with our biological double, we find we are literally beside ourselves.

Insofar as the subject is faced with the prospect of another that both is and is not itself, the clone represents the ultimate doppelgänger. Like the photograph described so poignantly by Barthes in *Camera Lucida*, the clone faces the subject with a technological image of itself from the outside or, equally if not more disturbing, with the sound of its own voice from the outside. This figure invokes a simultaneous recognition and misrecognition of this technically mediated imitation of ourselves, like the image we see of ourselves in a photograph, as Barthes describes:

In front of the lens, I am at the same time: the one I think I am, the one I
want others to think I am, the one the photographer thinks I am, and the
one he makes use of to exhibit his art . . . I do not stop imitating myself,
and because of this, each time I am (or let myself be) photographed, I inva-
riably suffer from a sensation of inauthenticity, sometimes of imposture . . .
I am neither subject nor object but a subject who feels he is becoming an
object. (Barthes 1981, 41)

These processes of psychic splitting in dialogic exchange with mediated others capture the extent to which any conceptualization of the
disturbance presented by the clone necessarily builds upon preexisting
imitations in the subject's unconscious constitution. The photograph
presents the viewer with a feeling of imposture because it reiterates
a foundational imitation of the self. The sense of self-imitation is produced at the moment when this feeling is transformed into an image.
The clone similarly provides a visualization of self-doubling. By analogy, we might paraphrase Barthes for a moment and say: faced with
my clone, I am at the same time: the one I think I am, the one others
(and now I) perceive myself to be, the one the scientist wants to show
to the world to exhibit his techniques. However, I am no longer the one
at all, but through genetic engineering have become more than one and
less than two.[1] So perhaps it should be: I am one of the two, whom
I thought was one; I am one of the two, both of whom I perceive as
other; I am half the equation the scientist wants to show to the world as
more than one but not fully two. But is it harder to imagine the clone
of another—of someone we love or repudiate, or of a relative or old
friend—than it is to imagine a clone of ourselves? Imagining the clones
of others presents another trajectory of interference played across numerous cultural sites: how are we to recognize the original and the copy
in the other; and how would the doubling of the object of our attachment disturb the singularity of its affective direction? Is this the anxiety
driving the genetic imaginary?

Of course, the analogy with the photograph is limited since the clone
figure is not (or not only) an image but rather presents the possibility
of concretizing imitation through biological corporeality. As such, the
clone both presents the subject with an image of itself from the outside
and confronts us with the depth of physicality to which a photograph
can only refer, as it displays a presence once there but already lost. Yet
the prospect of a biological clone is also me and not me (or you and not

you) in the way the photograph is and is not me, according to Barthes. The clone as the double thus involves both corporeality and its absence. These moves invoke the uncanniness of identical twins, who, as precursors of the clone, embody just such impossible paradoxes. But now identical twins are no longer the exception that proves the rule of human singularity, for the clone promises the possibility of reversing the laws of nature more generally.

Both figures (the identical twin and the clone) disturb the self-other relations of the subject's narcissism. Taken as a fundamental aspect of psychic development, rather than as a pejorative ascription (though we might note that, not insignificantly, narcissism has been disproportionately associated with homosexuality and femininity),[2] Freud's theory of narcissism explores its primary and secondary formations in erotic attachments to both self and others (ego instincts and object instincts). In the projections and introjections that characterize libidinal attachments, overvaluations of the other typically flow from a diminished sense of self. As Freud puts it, "another person's narcissism has a great attraction for those who have renounced part of their own narcissism" (Freud 1991, 82–83). We might wonder how to think about narcissistic attachments in the context of cloning, given Freud's categorizations:

A person may love: —
(1) According to the narcissistic type:
(a) what he himself is (i.e., himself),
(b) what he himself was,
(c) what he himself would like to be,
(d) someone who was once a part of himself. (Ibid., 84)

Slipping between Freud's subcategories of narcissism, the clone meets all and none of these criteria: both self and image in one, it offers a projected fantasy of the impossible object of narcissistic desire in a fleshly form. The clone figure appears to biologize psychic impulses (insofar as the self might be imagined to be embodied in another), and it thus presents a possible libidinal object of love or repudiation. We might finally be faced with our desirability, or lack of it, and with the nature of our own object choice. To the extent that the clone represents both self and image, it invites a reading of narcissistic attachment. Perhaps it should be thought of as the product of techno-scientific narcissism: the outcome of a project based on the desire to push beyond the limits of the human and into the realm of the divine. Taking nature to divine excess,[3] cloning has come to symbolize the multiple transgression of scientific iconoclasm.

If our fascination with images of the human form stems from our psychic inability to grasp the totality of our own physical presence, and if our attachment to photographs rehearses the phenomenological impossibility of seeing ourselves from the outside while experiencing being ourselves in the world, then perhaps the clone is only the genetic manifestation of the mutability of boundaries between self and other, inside and outside, and subject and object—which were always illusory oppositional constructions anyway. The clone confounds the distinction between original and copy upon which notions of imitation have depended. As an identical copy, it reveals the duplicity that already constitutes us. In other words, perhaps cloning represents the literalization of the fascination our own duplicity holds for us from the beginning. Geneticizing the human body through cloning means turning it into a reproducible code that reveals the vanity at the heart of the subject's self-imaging.

But duplicity is hardly a human equalizer. Some bodies, of course, have been marked as more duplicitous than others; some are better mimics, for they have had more practice. The history of imitation reveals who has authority to demand mimicry in others. The clone today belongs to an imaginary landscape of mimetic figures embedded within intersecting cultures of domination and regulation, in a world where some subjects have been required to mimic themselves endlessly for the benefit of others, or rather to mimic the version of themselves desired by others. Western modernity has been characterized by modes of repetition in which the privileged few have announced themselves as the bearers of universal subjectivity: it is their originality, their sense of generative power which must be shored up through the mimicry of others. Thus, the clone may be an all-too-familiar figure to those for whom artifice has always been a required performance of their subjection.

To mimic can mean to mime, and to mime can mean to act (OED, definitions 2, 3); thus, the art of imitation brings with it questions of audience and readership. The politics of this dialogical subject raises the question of who is performing what for whom (see also chapter 3). As part of a wide debate within feminist theory, Luce Irigaray has borrowed the word "mimesis" from philosophy to describe a strategy which might enact, in Mary Ann Doane's words, a "defamiliarized version of femininity" in order to shatter the specular economy of patriarchal relations (Doane quoted in Schor 1995, 53). Irigaray writes:

> One must assume the feminine role deliberately which means already to convert a form of subordination into an affirmation, and thus begin to thwart it . . . To play with mimesis is thus, for a woman, to try to recover the place of her exploitation by discourse, without allowing herself to be simply reduced to it. It means to resubmit herself—inasmuch as she is on the side of the "perceptible," of "matter"—to "ideas," in particular to ideas about herself that are elaborated in/by a masculine logic, but so as to remain invisible: the cover-up of a possible operation of the feminine in language. It also means to "unveil" the fact that, if women are such good mimics, it is because they are not simply resorbed in this function. *They also remain elsewhere.* (Irigaray quoted in Schor 1995, 53, emphasis in the original)

In playing the construction of woman's otherness back against its origins in the desires structuring patriarchal culture, Irigaray's strategy is to amplify the discourses of misogyny so that woman's masquerade might perform itself in such a way as to expose the "specular logic of phallocentrism" (ibid., 52). This strategy exaggerates what Naomi Schor calls "the bogus difference of misogyny" in order to transform patriarchal logic into a parody of its own absurd paradoxes (ibid., 53). As a mimic, the woman is required to perform this projection of herself as other to the masculine subject; but as this other (matter, body, object of desire), she is also an extension of the same masculinity. Schor synthesizes the multiple theories of feminine duplicity thus: first, in the context of feminism, there is the so-called masquerade, or the "old mimesis," which "names women's alleged talents at parroting the master's discourse" (including misogyny); second, "mimesis signifies not a deluded masquerade, but a canny mimicry," as "parroting becomes parody"; and third, "mimesis comes to signify the difference as positivity, a joyful reappropriation of the attributes of the other that is not in any way to be confused with a mere reversal of the existing phallocentric distribution of power" (ibid., 54). These three registers of mimesis implicate the enactment of femininity very differently, in ways which foreground the dialogics of such strategies: who is producing which version of woman for whom? Is the performer's intention caught by the audience in this slippery exchange of meanings, where the rehearsal of form is the point of the performance? If femininity is formal in this way, how should it be read across the relays of originals and copies at stake here? Does exposing its artifice reflect as much on its origins in masculine desire as it does upon the duplicity of the feminine (see also chapter 3)?

To place the new possibilities of human cloning within the history of feminist debate about mimicry is to theorize genetic simulation in terms of the slipperiness of these imitative relations. A psychoanalytic reading of cloning as biomimicry might be combined with a feminist critique of the investments of the masculine psyche in scientific knowledge production. Insofar as mimicry might be a strategic form of parodic playback (or even pastiche—see Dyer 2007), exposing the specular wish that the other reflect gratifyingly upon the ego of the original, perhaps the figure of the clone threatens to expose the sexual desires of techno-scientific cultures. In this sense, the production of a copy does more than just duplicate; it also transforms the original by facing it with its own desires and presenting it with an image of its own (diminished) embodiment. As Katherine Hayles (1999) has argued in relation to information technologies, fantasies of downloading intelligence into machines (the brain into the computer, for example) reproduce masculine visions of the agency and potency of the liberal humanist subject. Just as the desire of the masculine original for the same exposes the projection of otherness onto the woman through mimetic strategy, so genetic duplication brings to the surface the masculine subject's fascination with the self-same, which he projects onto an other as always more than himself. The clone figure is a mimic of the original insofar as it substantiates the notion of the same as other, and the other as the same. But in a culture where these mimetic dynamics have such a gendered history, how is masculine desire implicated in cloning scenarios? Does such duplicity feminize the desire of the masculine subject, or perhaps produce a feminine relation between masculinities? Does it expose his desire for difference as really a desire for the same (see also chapter 5)? How does the presence of the clone destabilize the authenticity and singularity of the masculine subject's desire? Determining what is threatening (and desirable) about the clone requires us to read mimesis as a necessarily dialogic set of exchanges between self and other in which the subject's deluded ambitions are literalized, and potentially exposed, in this external corporeal form.

Here the figure of the clone as biomimetic returns us to Freud's theory of narcissism, implicated in these later feminist conceptualizations of mimicry, and to the question of how sexual difference informs early psychoanalytic formulations. Freud suggests that "complete object-love of the attachment type" is typically, if not exclusively, male, and that it displays the "marked sexual overvaluation" corresponding to the transference of the "child's original narcissism . . . to the sexual object." This masculine overattachment is "traceable to an impoverishment of the ego as regards libido in favour of the love-object." In contrast, classic feminine narcissism is "unfavourable to the development of a true object-choice with its accompanying overvaluation" and is typically an "intensification of the original narcissism"; consequently, only women love themselves (and long to be loved by others) with such force (Freud 1991, 82). In Irigaray's rewritings of constructions of feminine narcissism discussed above, the place of the idealized woman as projected fantasy of the masculine self—that is, as representing sameness, and not the apparently marked otherness of difference—is critical. Here again, something masquerades as its opposite. Perhaps the clone figure might undo such binaries and face the masculine subject with his own narcissism, reflecting it back onto its source and exposing the depletions of the original and his imposture as it does so. Like the classic mirror-stage encounter, the figure of the clone represents the subject to himself as troublingly less than he imagined himself to be. Embodying an externalized image of coherence and a totality that the subject cannot feel internally, the clone contradicts the idealized self-image in the subject's imaginary past: the clone figure robs the subject of his fantasy self by substantiating its necessarily inferior form (the corporeal contradiction of his imaginary ideal ego). More than this, in its spatial disturbance, the clone requires a multiple identification—an identification from the outside, from the one who defines himself through his autonomous singularity. We might say that in being faced with his cloned other, the masculine subject is thus required to recognize himself in what he always imagined might be greater, but suddenly perceives to be less, than he thought he was. Yet in using this formulation, we fail to speak of the clone's own subjectivity which, presumably, simultaneously doubles and replays this dynamic.

The figure of the clone thus steps into a long history of ambivalence in which the dominant subject has required others to fulfill his fantasy projections through mimicry while disavowing his own investment in such a performance. As Homi Bhabha suggests in his famous account of mimicry in the colonial project:

> Colonial mimicry is the desire for a reformed, recognizable Other, *as a subject of a difference that is almost the same, but not quite.* Which is to say that the discourse of mimicry is constituted around *ambivalence*; in order to be effective, mimicry must continually produce its slippage, its excess, its difference. The authority of that mode of colonial discourse that I have called mimicry is therefore stricken by an indeterminacy: mimicry emerges as the representation of a difference that is itself a process of disavowal. Mimicry is thus the sign of double articulation. (Bhabha 2005, 85, emphasis in original)

It is precisely the indeterminacy that the neuroses of the colonial subject puts into play which then return to haunt his desire for ultimate authority. This is the "double command" of colonial authority—"be like me, don't be like me, be mimetically identical, be totally other"—identified by Diana Fuss in her discussion of Frantz Fanon's work on psychoanalysis and colonialism (Fuss 1994, 23). According to Fuss, this impossible imperative places the colonial other "somewhere between difference and similitude at the vanishing point of subjectivity," as the "colonized are constrained to impersonate the image the colonizer offers them of themselves" (ibid., 23–24). If, as Fuss argues, "'white' defines itself through a powerful and illusory *fantasy* of escaping the exclusionary practices of psychical identity formation," then might cloning be seen as the technical extension of such omnipotent white fantasies (ibid., 24, emphasis in the original)? As Fuss suggests, psychoanalytic theories of colonialism have described how the colonized subject is enjoined to "identify and disidentify simultaneously with the same object, to assimilate but not to incorporate, to approximate but not to displace"; ultimately this is an "injunction *to mime* alterity" (ibid., emphasis in the original).[4] And yet this risky strategy contains the seeds of its own potential implosion. Although postcolonial theorists such as Bhabha (2005) use mimicry to name a strategy of colonial domination—indeed, they identify mimicry as one of the most "elusive and effective strategies of colonial power and knowledge," as an "emphatic instrument of political regulation, social discipline and psychological depersonalisation"—Fuss argues that the distinction between a mimicry of subversion and one of subjugation is frequently hard to make, precisely because of the internal contradictions of the desire underpinning such imperatives (Fuss 1994, 24).

Cloning might disturb our psychic horizons because it threatens to literalize the desires behind these continuing, mimetic projects. The prospect of the clone gives a troublingly fleshly presence to the desire for the self-same, facing the dominant subject with his own idealizations and with his own limits. Human cloning represents a form of biomimicry in the senses of both imitation and shared corporeality. It confounds the colonial desire to hold fast the distinction between *Innenwelt* and *Umwelt* through which, as Fanon puts it, Martinique is the Umwelt to Europe's Innenwelt, since it places the subject's double within the Umwelt normally reserved for the mimicry of the other. Once again the spatial relations through which the subject has stabilized his sense of self are undone. The distinction between the inner self and the environment, so pivotal for the colonial regimes of othering, breaks down when the subject is confronted with the same as other in the Umwelt of his fantasies.

The clone figure rehearses previous disavowals upon which the authority and singularity of the subject's domination has been based. If the mimicry at play in colonialism concerns anxieties about racial purity and about the threat of the other within, the clone enters the scene as a product of science haunted by its eugenic past, in which purity and mixtures were a matter of life and death (see the introduction and chapter 6). In some ways, cloning promises a strange ideal of purity. Avoiding the unpredictable mixing of heterosexual procreation, cloning offers a fantasy of pure replication from a singular source. The messiness of reproduction (not to mention the abject scene of birth itself) is replaced by the hygienic, clinical control of the laboratory: cloning as purified biology. Yet in other ways, the figure of the clone frequently represents the new possibilities of transgenic hybridity, extending the threat of miscegenation that has haunted white culture to new kinds of mixing and to cross-species combinations. Now genetic information can be transferred between bodies, and cell parts can be blended across traditional boundaries between people and even between species. In the genetic imaginary, nature becomes a mutable sign as the clone figure seems to represent both more of it and less of it in its racialized embodiments: white bodies translate into artifice; black bodies become naturalized; and new biomixtures connote a lost fantasy of racialized purity. A biology of genetic interference thus makes visible some of the psychic legacies of colonial mimicry.

If the clone threatens to face the subject with his own psychic disavowals and repressions, then both the image and the body are simultaneously at stake in biomimicry. Presenting the self with its status as image in that moment of genetic duplication, cloning unites modes of image production and of corporeal enactment. As I discuss in chapter 5, Mary Ann Doane draws on a famous 1929 essay by the psychoanalyst Joan Riviere to argue that it is not mimicry but the notion of masquerade which is more useful in theorizing the marking of particular bodies as imagistic (Doane 1991a, 25). In a much-quoted passage, Riviere takes up the problem of the applicability of her theory beyond the particular patient (a female academic) in her case study:

> Womanliness therefore could be assumed and worn as a mask, both to hide the possession of masculinity and to avert the reprisals expected if she was found to possess it—much as a thief will turn out his pockets and ask to be searched to prove he has not the stolen goods. The reader may now ask how I define womanliness or where I draw the line between genuine womanliness and the "masquerade." My suggestion is not, however, that there is any such difference; whether radical or superficial they are the same thing. (Riviere 1986, 38)

For Doane, "the theorization of femininity as masquerade" offers feminists "a way of appropriating this necessary distance or gap, in the operation of semiotics systems, of deploying it for women, of reading femininity differently." If femininity is the "play of masks" for Riviere, Doane argues, and yet "there is no censure involved in claiming that the woman hides behind the mask when the mask is all there is," then the contradictions within such a theory potentially de-essentialize woman and her place as image in patriarchal culture (Doane 1991b, 37). Unlike mimicry—which here suggests the conscious adoption of a strategic and ironic playfulness in relation to misogynist language practices, in order to enact patriarchy's ascription of dubious otherness to women— masquerade, for Doane, is an unconscious display of the accouterments

of femininity in a compensatory gesture (note the quite different connotations from Shor's and Irigaray's use of the term above). For Doane,
masquerade suggests a "mode of being for the other" which might
destabilize the image by challenging the relation between lack and absence in Lacan's theory of language and in his theory of sexual difference (ibid., 33–35). The problem in both psychoanalysis and the cinema,
Doane claims, is that femininity is defined as closeness or proximity.
Masquerade, she suggests, might be deployed to explain the constraints
of the place of "woman as image" within a sign system governed by the
law of the paternal signifier (ibid., 35). Analyzing the complex relationship between femininity and the image, and between performing the
desire of the other and visual display, she claims that masquerade might
escape the ascription of femininity as nearness because of the "curious
blend of activity and passivity in masquerade . . . and the corresponding blurring of the opposition between production and reception"
(ibid., 39). Offering a way to find the contradiction within the psychoanalytic theory itself, masquerade, for Doane, "attributes to the woman
the distance, alienation, and divisiveness of self" required to refuse to
read femininity conventionally (ibid., 37).

Here two interrelated points emerge which are central to thinking
about the figure of the clone: first, masquerade, like cloning, is a form of
what we might call intercorporeal dialogics, which depend upon readership and spectatorship; and second, masquerade, like cloning, suggests deception and even criminal associations. The first point relates
to the dialogics of identity discussed in chapter 3, the ways in which
identities are not simply constituted, embodied, or lived out by subjects
themselves but are also always read (typically seen) by others. Cloning
simultaneously faces us with absolute certainty and profound uncertainty about who we really are. This first point, the new forms of literacy
required to render the geneticized body legible, leads inevitably to the
second point, that of the slippage between duplication and duplicity
(which haunts all the theories of imitation discussed so far). Cloning is
a form of duplication in which identity theft is made visible. The crime
against nature is elided with criminal intentions to deceive: both involve duplicity; both involve deception. The clone thus belongs to a long
history of imagined bodies whose deceptive surfaces conceal the secrets of hidden depths.[5] If womanliness as masquerade involves a guilty
coverup (much like the thief who turns out his pockets), then cloning

moves criminal intent center stage. Albeit unconsciously, womanliness as masquerade is like a decoy, an attempt to throw potential aggressors off the scent of the desired theft of masculine privileges and an avoidance of punishment. Its success depends upon the audience's reading the surface performance and seeking no deeper evidence of innocence or authenticity. In other words, it depends upon reading the body of the woman as image while she simultaneously performs herself as intellectual. This discrepancy between the image and the woman's desire offers the gap upon which Doane pins so much hope. Here lingers the promise of a distance typically denied the woman. Likewise, the clone becomes an impostor whose future depends upon convincing his audience of his legitimacy or authenticity. The distinction between mimicry and masquerade, as Fuss puts it, "depends upon the degree and readability of its excess" (Fuss 1994, 24). The prospect of cloning plays upon our fears about genetic illiteracy. The indeterminacy of the embodied subject and the impossible certainty of its legible surface thwart our ceaseless desire for certainty and security. Like the femme fatale, whose treacherous deceptions have produced an endless fascination with reading the discrepancies between her seductive surface image and her hidden evil motivations, the clone figure requires us to sharpen our skills of fraud detection (see chapter 5).

The figure of the clone brings together duplication and duplicity in our very modes of embodiment. Here we slide from imitative forms of mimicry into duplicitous modes of imposture and impersonation. As I shall elaborate in the following chapter, the concept of impersonation better captures the intercorporeality of cloning and the shared habitation of genetic embodiment, but most importantly, it condenses imitation and deception. The impossibility of fully grasping the idea of cloning reinforces our desire to hold these apart. Questions of passing lie at the heart of this slippage, as Carole-Ann Tyler (2003) has argued. Cloning involves multiple forms of passing: passing as, passing on, passing across, and passing through. Defying the usual assumptions about the authentic and the inauthentic, the clone simultaneously passes as itself and, in undoing notions of originality, passes over the self.

In her reconceptualization of gender as performative rather than expressive, as something one does rather than something one is or has, Judith Butler (1990) famously reverses the traditional teleologies of originals and copies. Challenging the conventional move to cast lesbian and gay sexualities as poor imitations of a heterosexual original, she asks what might it look like if "the normal" or "the original" (here, heterosexuality) were to be revealed to be already a copy, and "an inevitably failed one, an ideal that no one *can* embody," an original that "all along . . . was derived" (Butler 1990, 139, emphasis in the original). If gender parody did not assume an original as its object—if instead the parody is "*of* the very notion of the original"—then for Butler, the parody might reveal "that the original identity after which gender fashions itself is an imitation without an origin . . . which . . . in its effect—postures as an imitation" (ibid., 138, emphasis in the original). With all its connotations of gay male subculture, the figure of the clone brings the imitative imperatives of gender and sexuality together with scientific desires to copy nature. Just as heterosexuality for Butler is already a failed copy of itself, so the clone represents a doubling that reflects back on the body's claims to authenticity and originality. But more than just a visual doubling that destabilizes the original-copy dualism, the figure of the clone also gives the most visible form to the biological body as genetic code. We might thus ask, with Butler, "in what language is 'inner space' figured," and "how does a body figure on its surface the very invisibility of its hidden depth?" (ibid., 134). Extending Butler's questions, we might consider how the clone produces a sense of the new hidden depths of genetic embodiment: how does this visible doubling reconfigure the body's deeper truths (see chapter 2)?

One way to begin to answer these questions is to think about how the surface sameness of the clone constitutes the geneticized body beneath as unoriginal. In its newly coded form, the body becomes an assemblage of genealogical information—merely the carrier for genes across generations. Combining and condensing genetic information from two biological parents, the offspring body becomes the two in one whose inheritance can be traced in minute linear detail. What we might call the informationalization of the body as genetic code thus both promises absolute verification of its biological origins and turns it into a posthuman copy, a blending of previous incarnations. In this sense, the more the body can be authenticated through genetic information, the less original it becomes. The figure of the clone produces this paradoxical vision: identical to its copy, the original body gives up its claims to singularity and, by virtue of being duplicable, begins to relinquish its sense of authentic originality.

Thinking about cloning as biomimicry in a book on the cinematic life of the gene is, in part, a way to consider the coincidence of the imitative techniques of science and the cinema, but it is more than this. It is also an attempt to explore the conceptual basis of the shared imaginative history of the body and the image. Just as theories of masquerade in film have played upon the idea of the original as itself being an imposture, so too we might conceive of the clone as a visualization of how the human body appears increasingly unoriginal.

[5] Genetic Impersonation and the Improvisation of Kinship

Gattaca's Queer Visions

In the opening sequence of *Gattaca* (Andrew Niccol, 1997), an enigmatic scene of minimalist formal beauty gradually becomes a display of the shedding of abject bodily detritus for the purpose of an elaborate disguise. In the first few shots of the film, the excessive visual magnification of nail clippings, strands of hair, and flakes of skin effects a visual deception upon the audience: the nails look like large, crescent-shaped pieces of frosted glass, the hair like lengths of rubber piping, and the shower of skin like a beautiful snowfall. Initially unidentifiable, these gigantic bodily fragments fall in slow motion, hitting the ground with a thudding vibration as they eventually settle on a luminous blue surface that fills the screen. The exquisite visual poise of the mise-en-scène is underscored by Michael Nyman's melancholic, minimalist music, which completes the seductive aesthetic of tranquility and perfection governing the scene. As the magnification lessens, the human

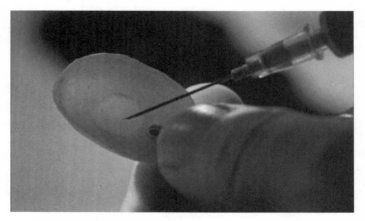

15. Vincent (Ethan Hawke) injects Jerome's blood into a
false fingertip in *Gattaca* (Andrew Niccol, 1997).

source of these falling objects is slowly revealed: a chin is being shaved;
a rather androgynous chest is being scrubbed; muscular arms are being
abraded. A male figure is shown vigorously and painfully discarding
all these external bodily traces in a cubicle bathed in deep blue light;
he leaves the shower room, and, with a gesture whose ease suggests
daily habit, turns on the incinerator inside it to obliterate all evidence
of his physical presence. This ritual cleansing then starts to look like
something else: the work of disguise. The man takes two medical infusion bags from a refrigerator; he attaches one to his upper leg at the
level usually reserved for a suspender belt and injects blood from the
other into a pocket in a false fingertip (see figure 15), which he sticks
on the index finger of his right hand, using an eyebrow brush to secure
the edges. His routine complete, he drives to work at Gattaca, a space
exploration company.

Gattaca is a film about a genetically defective man, Vincent (Ethan
Hawke), who disguises himself as a genetically perfect one, Jerome
(Jude Law), as I discussed briefly at the beginning of the introduction
to this book.[1] The film's opening scene offers a disorienting visual deceit
that encapsulates the film's more general exploration of the artifice of
masculine perfection in the age of genetic engineering. These hyper-
magnified shots of bodily fragments foreshadow the film's more general
investigation of the relationship between scientific and cinematic technologies. The scientific and the cinematic gazes merge as microscopic

magnification and extreme close-up shots combine in this credit sequence to bring us so close to these corporeal fragments that we feel
as if we could almost touch them.[2] Seeing the body so close up and
yet misreading it so profoundly invites contemplation of the relationship between seeing and knowing, between observable corporeal surfaces and the identities beneath them. Should we trust the knowledge
promised by visual technologies? To what extent can they deliver the
truth of a person's identity? If these close-up shots offer the promise of
truth through the magnified and slowed-down image, ultimately they
do so only to undermine its certainty: more vision, in this film, often
leads to less knowledge, to distortion, misreading, and even a reversal
of perception.[3] What appears under the microscopic cinematic gaze as
a beautiful snow scene is transformed into a shower of discarded, dead
skin by a slight shift in visual perspective. The slowing down of time
combines with the magnification of objects to transform bodily abjection into aesthetically pleasing abstraction. And as the whole trajectory
of the film demonstrates, visual evidence and genetic evidence are not
seamless equivalents, nor are they reliably transparent: both are open
to manipulation and susceptible to the indeterminacies of interpretation. Even where genetic engineering promises to make biology definitively predictive, image and identity cannot simply be read off from
technology; they must be achieved through it. As in *Species* (Roger
Donaldson, 1995), identity here is informatic; it is a question of convincing your audience that the genetic screening technologies designed
to return the sign of the body to transparency make human interpretation unnecessary.

Set in the not-too-distant future, *Gattaca* is a dystopian science-
fiction film in which a rigidly hierarchical regime of genetic screening
and selection brutally govern the fate of individuals. A person's DNA
can be tested from a licked envelope, a shaken hand, a kissed mouth,
or a shed eyelash. For a fee, anyone can use a strand of hair to have a
potential sexual partner's DNA sequenced in a matter of minutes, to
test for genetic compatibility. This is a world in which the blood test has
replaced the police interrogation, and the urine test has replaced the
job interview. The inequalities that result from these genetic selections
are presented most sharply in Gattaca, the workplace of the same
name. There, valids (those selected from genetically superior embryos)
occupy the high-status positions and are valued for their exceptional

intellectual and physical attributes, while those not selected, the in-
valids, a term with obvious connotations of physical inferiority, are
cleaners who service the building and its employees. *Gattaca* presents
the nightmare of a new form of segregated workforce whose classification seems to have made perversely irrelevant the traditional antidemo-
cratic hierarchies of race, class, and gender: in this world of genetic
normativity, even white, middle-class men like Vincent can be destined
for repetitive, menial labor.[4] Motivated by sibling rivalry, however, he
proves his brother, his father, and the scientists wrong, but his mother
right, defying the limits of his genetically predicted future by becoming
a space navigator at Gattaca. To achieve this rebellious end, he adopts
the genetic identity of a valid—Jerome Eugene Morrow, an Olympic-
standard swimmer whose genetic code is practically perfect but whose
accident abroad has left him wheelchair bound (see the introduction).

The quest for genetic perfection that governs *Gattaca*'s dystopian
fantasy is articulated through the film's mise-en-scène of formal repetition and sterility (see figure 16). As David Kirby has written, "visually,
GATTACA conveys an antiseptic world that has been purged of imper-
fections . . . [The sets] show a sterile and blemish-free world filled with
smooth stainless steel surfaces" (Kirby 2000, 204). The preoccupation
with visual perfection within a world of genetic normativity is centrally
elaborated through Vincent's disguise as Jerome. Vincent imitates visual perfection by impersonating Jerome (whose embodiment of desirability is amplified by the casting of Jude Law) and by displaying the
exacting precision of disguise necessary to avoid genetic detection.
The opening scene shows Vincent's daily physical transformation into
Jerome as a spectacle of identity production. Yet in so doing, it displays
for the cinema audience precisely the deceit that genetic screening
promises to eradicate. In order to imitate Jerome's genetic perfection,
Vincent changes both his outward appearance to approximate Jerome's
image and his so-called informatic code by substituting the markers
of Jerome's genetic information (blood, urine, skin, and hair) for his
own. In this spectacle, the technologies of imitation are set against the
technologies of genetic testing. With Jerome's assistance, Vincent becomes master of the image and of deception. Artifice is their greatest
ally against genetic determinism.

If the veracity of visual evidence is destabilized in *Gattaca*, so too
is the veracity of the body as the guarantor of the apparent truths of
gender, genealogy, and kinship. In placing Vincent and Jerome's crimi-

16. Sequence, order, and control govern the mise-en-scène of the gymnasium at the aerospace firm in *Gattaca* (Andrew Niccol, 1997).

nal deception at the heart of the discrepancy between visual evidence and genetic evidence (between the image and information), the film arguably undoes the singularity of masculine sovereignty and queers traditional forms of kinship as much as it does conventional forms of vision.[5] Biological inheritance is displaced through the rejection of genetic normativity, as the audience gradually becomes privy to the alternative bonds of relatedness and forms of intimacy between the two men. The production of Vincent and Jerome's shared bodily substances requires the improvisation of an intermasculine kinship with a distinctly queer feel.[6] In their reinvention of the blood tie as genetic impersonation, Vincent and Jerome unite against normative injustice in their new form of shared embodiment.[7] Like the male "couples" in *Rope* (Alfred Hitchcock, 1948) and *Swoon* (Tom Kalin, 1992), Vincent and Jerome are locked into a mutual dependence that is sometimes claustrophobic: their combined ingenuity is required to execute their crime successfully. But unlike their predecessors in this relationship, Vincent and Jerome are tied to each other through a commitment to a new, fabricated persona they share (one gives his body, the other his dream), and thus their loyalty to each other is always also a loyalty to themselves. Their embodied deception defies both the conventions of masculine singularity and the laws of genetic determinism through a reconceptualization of identity beyond traditional definitions of genealogy, gender, and heterosexuality.[8]

In this sense, although *Gattaca* is not strictly a cloning film, like *Multiplicity* (Harold Ramis, 1996), *Alien: Resurrection* (Jean-Pierre

Jeunet, 1997), *The 6th Day* (Roger Spottiswoode, 2000), *Star Wars: Episode II—Attack of the Clones* (George Lucas, 2002), or *The Island* (Michael Bay, 2005), it nevertheless shares the genre's preoccupation with technologies of duplication and fascination with deviant forms of relatedness, discussed in chapter 2. Rejecting the injustice of genetic determinism, *Gattaca* stages a cinematic vision of cloning through the imitative replication of masculine perfection. Vincent arguably becomes Jerome's clone insofar as he successfully passes as a previous version of him at work. The narrative tensions around passing (will Vincent be caught?) combine the familiar concerns of science fiction about the problem of the authenticity of identity with the suspense around detection.[9] The film also reiterates a set of connotations around the secrecy at stake in the homosexual closet. In the face of their possible discovery, Vincent and Jerome's mutual loyalty and devotion to their secret commitment is repeatedly tested and ultimately strengthened. The intimate exchanges between the two men (of fluids, knowledge, dreams, and identities) expose the limits of their individual autonomy and push to the surface the mutuality of their improvised co-embodiment. In "cloning" Jerome, the two men enter the terrain not only of queer kinship but also of homoeroticism. With their obvious homosexual connotations, passing and cloning both give queer implications to the criminal intimacy of genetic deception in the film.

In this chapter, I focus on the ways in which these various queer improvisations are set in tension with the more traditional cinematic organization of the desire for heterosexual difference. In particular, I am interested in the combination of the display of the artifice of a masculine impersonation with the queering of kinship in a postgenealogical age (see Butler 2000 and Franklin and McKinnon 2001). Although Vincent's disguise requires a consideration of the operations of sexual difference in this spectacle of masculine artifice, the intimacy of his collaboration with Jerome as master of genetic disguise calls for queer theory to unravel the homoeroticism of impersonation and new loyalties of queer kinship. Taking *Gattaca* as my central text, I examine the problem of rendering masculinity an authentic, stable identity when it is produced as a technological achievement of genetic disguise jointly authored by two male collaborators.[10]

The display of the body as artifice in the cinema has been widely debated within feminist film theory through the concept of the masquerade, as

I discussed in the previous chapter.[11] Here, I return to these debates to consider the place of masculinity within theories of the masquerade. Drawing on psychoanalysis, feminists have claimed that the masquerade is not only closely connected to femininity but is also inextricable from its cultural ascription within patriarchal representational systems such as Hollywood cinema. According to these analyses, masculinity can have a place only outside the performance of the masquerade.[12] As chapter 4 detailed, Mary Ann Doane (1991a and b) has returned to the psychoanalyst Joan Riviere's 1929 case to argue that construing femininity as a masquerade might provide a means to explain the constraints of the place of woman as image within a sign system governed by the law of the paternal signifier. Riviere suggests that like "homosexual men [who] exaggerate their heterosexuality as a 'defence' against their homosexuality . . . women who wish for masculinity may put on a mask of womanliness to avert anxiety and the retribution feared from men" (Riviere 1986, 35). Importantly, Doane argues, "Riviere's patient actively strives to produce herself as the passive image of male attention" (Doane 1991b, 39). Doane emphasizes here how central reading and interpretation are to understanding both masquerade and femininity. Although she sees the masquerade as "haunted by a masculine standard, masculinity as measure is not internal to the concept itself"; rather, she argues, "in masquerade, masculinity is present as the context provoking the patient's reaction-formation" (ibid.). Masculinity's relationship to masquerade is thus placed outside the concept itself and takes form only in producing the reaction in the audience or the readership.[13]

If duplicity through artifice has had such a strongly feminine set of connotations within feminist film theory, how might we interpret the masculine imitation of masculinity in a film such as *Gattaca*, in which two male characters combine their ingenuity and resources to produce a deceptive image of genetic perfection?[14] In the opening scene discussed above, a spectacle of fetishistic bodily cleanliness and grooming, such attention to the detail of a man's bodily transformation establishes an immediate association with femininity: that is, an identity achieved through the labor of producing a perfect body using whatever artifice is available, and bearing whatever pain is necessary. In this scene, a number of close-up shots of Vincent's body confirm these more generally feminine connotations: the smooth, hairless skin; the somewhat androgynous chest; the positioning and attachment of the urine bag as if it were a garter belt; and the delicate and expert use of the eyebrow brush. Moreover, there is a deeper association here based on the troubling potential for disguise in an identity premised on artifice. If the work of femininity is not only to produce an image but thereby to achieve an identity, the figure of the woman can be considered a site for endless suspicion. Taking its most treacherous form in the figure of the femme fatale, the deployment of a physical image to effect a disguise is a move associated with the deceit of femininity. In this scene, the scrupulous shedding of hair and skin to produce a smooth, less legible, surface and the meticulous attachment of blood and urine from refrigerated infusion bags all imply that the end product of this ritual is some kind of deception.

Vincent's impersonation of Jerome is potentially feminizing not only as a prosthetic spectacle, but also in its clear intention as criminal disguise. As Doane points out, Riviere repeatedly associates the masquerade with theft and "stolen goods," since womanliness as masquerade covers the theft of masculinity (Doane 1991b, 34). In *Gattaca*, Vincent is not only associated with fraud; more significantly, his deception is narratively tied to a murder. The connection between Vincent and criminality is elaborated in a number of ways: the sight of the murder of one of Gattaca's directors coincides with the dissolve into an extended flashback, in which Vincent's voice-over explains the history of his fraudulent identity to the audience; as the murder investigation accelerates, detectives (headed, unbeknown to the audience, by Vincent's valid brother, Anton) and their genetic screening devices threaten the continued success of Vincent's disguise; and, his colleague and later lover, Irene (Uma Thurman), becomes suspicious that he did in fact murder the director. The exploration of Vincent's deception thus becomes narratively inextricable from the investigation of the murder. Vincent is structurally placed in the traditional position of the femme fatale— through its disguise, his body is potentially associated with treachery.

In imitating Jerome, Vincent has stolen a masculinity that is not his own. Does Vincent thereby somehow inhabit the traditionally feminized place of one who performs the image for the other? How is this spectacle gendered, if it displays an imitation of a better masculinity? In her original essay, Doane wrote:

> The masquerade, in flaunting femininity, holds it at a distance. Womanliness is a mask which can be worn or removed. The masquerade's resistance to patriarchal positioning would therefore lie in its denial of the

production of femininity as closeness, as presence-to-itself, as, precisely, imagistic . . . To masquerade is to manufacture a lack in the form of a certain distance between oneself and one's image. (Doane 1991a, 25–26)

If, as Judith Butler has argued, "Riviere's text offers a way to reconsider the question: What is masked by masquerade?" (Butler 1990, 53), it is tempting to read the production of Vincent's imitative masculinity in *Gattaca* as the reverse of Doane's description: producing a closeness to the image that masculinity usually lacks; closing the gap between the self and the image usually reserved for femininity; or as the feminizing mask that sutures image together with identity through genetic disguise.[15] But here the limits of the concept of masquerade become apparent: with their exclusively visual emphasis, theories of masquerade cannot capture the dynamics of genetic disguise at stake in *Gattaca*. Passing through masquerade, Vincent perfects not only the image but also the intelligibility of the body through the legibility of the code. Confirming the importance of Doane's emphasis on interpretation in the case of femininity as masquerade, yet moving beyond the image as the exclusive site of the production of identity, genetic disguise here confounds such formulations. Vincent's imitation of genetic perfection is not so much a feminizing masquerade as an informatic imposture that turns the meaning of identity inside out. Identity deceptively becomes legible through the transparency of the genetic code, as technologies of interpretation open up possibilities of interference and reinvention. Biogenetic mimesis becomes the new disavowal of authenticity.

This twist to the status of the image in *Gattaca* problematizes any straightforward reading of Vincent's disguise as simply a feminizing masquerade. Throughout the film, the promise of the truth of visual evidence is subtly undermined even as it is reiterated: close-up shots of fragments of the human body refer back to the credit sequence, where the microscopic modes of spectatorship produce a striking proximity to the image, but do so ambiguously. In a world in which everyone's genetic code can supposedly be technologically translated into a visual image (the genetic ID card shows the person's name, DNA sequence, and photograph), the authorities are so sure of the infallibility of their regulatory apparatus that no one actually looks at photographs any more. As the broker who introduces Vincent and Jerome says, "When was the last time anybody *looked* at a photograph?" (see figure 17). While intended for social regulation, the ubiquity of blended genetic

17. Vincent (Ethan
Hawke) makes himself
resemble a photograph
of Jerome (Jude Law) in
Gattaca (Andrew Nic-
col, 1997).

and surveillance technologies introduces new possibilities for disguise.
As Hillel Schwartz writes: "In a world of proliferant degrees and diplo-
mas, impostors have more room than ever to move on from one half-
life to the next. These days embossed papers substitute for personhood,
identification cards for identity, licences for learning" (Schwartz 1996,
71). The proliferation of information renders looking redundant. Turning their domestic apartment into a stylish cloning laboratory, Vincent
and Jerome succeed in an elaborate deception that reflects the blind
spots of the scientific gaze back to the corporate authorities at Gattaca,
dodging the scrutiny of the scientific gaze by hiding in the shadows of
its own occlusions.

The instability of the visual evidence of geneticized bodies in *Gattaca*
can be read as an analog to the film's exploration of masculine desire.
On the one hand, Vincent's highly conventional masculine drive and
ambition propel the narrative forward and are central to his success
in bringing satisfactory narrative closure. As the surveillance tightens,
the new Jerome becomes tougher and more determined: as Jerome,
he becomes invincible (not-Vincent). His potentially feminizing self-
fabrication operates in the service of his masculine desire for sovereignty and agency. On the other hand, there is also a challenge to the
stability of the all-seeing, all-knowing position of the masculine spectator, made famous by Laura Mulvey's (1989) critique of the visual
pleasures of narrative cinema. Each time we might feel seduced by

the fantasy of panoptic vision through the point of view we share with the protagonist, our omnipotent delusions are undermined—as Vincent himself faces the next unexpected challenge, and as we are dislodged from the security of such an alignment. For example, the surprising discovery that the detective in charge of the murder investigation is Vincent's younger brother reveals that our hero has been hiding something even from us. Moreover, although the film deploys typical Hollywood techniques (such as flashbacks and voice-overs) in order to position the spectator with Vincent, they are undermined by the lack of a sense of emotional depth to Vincent's interiority. Like the forms of its presentation, Vincent's subjectivity remains a convention.

Similarly, although the narrative is structured around Vincent's desire to achieve heroic status and to prove the justice of meritocracy over genetic determinism, the film's own awareness of this as a convention of masculinity confounds its straightforward alignment with the spectator's pleasure. Vincent's desire for autonomous agency is presented as precisely that—his desire—and it is never realized as his identity. As we shall see, Vincent never fully inhabits "valid" masculinity; he is never secure as the author of his new identity as Jerome. The film's repeated thwarting of Vincent's desire to author his own success is a narrative device to produce suspense, and yet it has the cumulative effect of questioning the foundational drive behind the desire. *Gattaca* explores Vincent's masculine drive and, in placing him outside the dominant eugenic values of his society, invites us to invest in the success of his deception; yet in denying him agency in his final achievement, the film ultimately leaves his desire for autonomy as a fantasy position. This works in conjunction with the way the film places him (along with the eugenic project of Gattaca) at one remove from the critical spectator position it establishes. Vincent wants to be accepted by the very institution whose values his story has taught us to mistrust and abhor, and thus our position in relation to his success remains partially skeptical. Like the breathtaking scale of the monumental architecture (see figure 18) or the impressive symmetry of the chrome interiors, Vincent's ambition belongs to a masculine order that the film renders politically dubious, based as it is on an illusion of control, predictability, and order. The film presents this fantasy of masculinity through a series of reiterative performances, and in so doing it both animates the masculine desire for omnipotence and disavows the credibility of these performances by

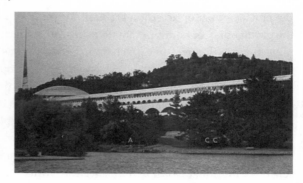

18. The counterpoint to microscopic genetic engineering: the monumental scale of Frank Lloyd Wright's Marin Civic Center used as exterior in *Gattaca* (Andrew Niccol, 1997).

locating them firmly within a eugenic aesthetic, associated with dangerous delusions of totalitarianism and fascism.

The modes of disjunctive temporality which operate in the film further mark the gap between Vincent's desire and that of the spectator: *Gattaca* is an all-too-familiar version of the future. Using clichéd fantasies of techno-scientific endeavor from the film's not-too-distant past (such as rocket science and space travel), *Gattaca* presents the masculine desire governing the hierarchies of a genetically determinist world as an ironic reflection on the modernist vision of the earlier period in which they are placed: the flashback to Vincent's childhood offers sepia-tinted scenes of stereotyped 1950s family life, together with retro-style furniture, cars, and clothes; the use of space exploration as the mise-en-scène of masculine ambition plays with a now outmoded notion of space travel as the final frontier for the progress of mankind. The image of a rocket being launched into space (with which the film closes, to symbolize the realization of Vincent's dream) places the spectator in the paradoxical temporality of being transported back to the future.

Vincent's masculinity in *Gattaca* is thus the rehearsal of a desire rather than the achievement of an identity. As Joan Copjec has pointed out, according to Lacan, "no man can boast that he embodies this thing—masculinity. . . . All pretensions of masculinity are then sheer imposture; just as every display of femininity is sheer masquerade" (Copjec 1994, 234). Vincent's genetic disguise makes him the impostor who exposes the more general facade of authentic masculinity. The imposture is as much about the repetition of the impossibility of masculinity (as the invincible, autonomous agent of events; the original and originator of meaning) as it is about the securing of its authentic form: "Impostures succeed because, not in spite, of their fictitiousness. They take wing with congenial cultural fantasies" (Schwartz 1996, 71). Vincent performs the pretence of masculinity, and his elaborate disguise is a cinematic enactment of the impossibility of ever embodying its literal form. In short, Vincent is an impostor whose disguise reveals him to be surrounded by impostors. For Copjec, Lacan's "desubstantialization of sex . . . has allowed us to perceive the fraudulence at the heart of every claim to positive sexual identity" (Copjec 1994, 234). Butler elaborates Lacan's position thus:

> [Lacan] poses the relation between the sexes in terms that reveal the speaking "I" as a masculinized effect of repression, one which postures as an autonomous and self-grounding subject, but whose very coherence is called into question by the sexual positions that it excludes in the process of identity formation . . . The masculine subject only *appears* to originate meaning and thereby to signify. His seemingly self-grounded autonomy attempts to conceal the repression which is both its ground and the perpetual possibility of its own ungrounding. (Butler 1990, 44–45, emphasis in the original)

In *Gattaca* the masculine subject "appears to originate" not only meaning, but also "life itself."[16] The repression required for the fraudulence of identity to succeed is turned into a science. The maternal body, as Susan George (2001) has argued, is displaced by the genetic selection of embryos before they are implanted into the woman's body, rendering the mother marginal to reproduction. Contrasting with the very physical scene of Vincent's birth earlier in the film, which shows the sweat and pain of his mother in labor, Anton's genetic selection shows four embryos pictured on a computer screen. Vincent's parents must agree to the suggested selection of the chosen one, the singular promise of embodied perfection. The geneticist (like the manager at the company who will later reject Vincent as an in-valid before the job interview begins) is black, suggesting the separation of eugenics from its racist past and cautioning against the new hierarchies beyond race that genetic normativity might produce. The display of the four embryos on the screen, alongside the recitation of the genetic information, transforms reproduction from a scene of human risk and adventure into one of an exact, predictable science, and conception into a disembodied virtual selection of a known entity. Reproduction thereby becomes a form of authorship, as the masculine subject becomes the originator of meaning, and science takes over paternity.[17]

The promise of genetic screening is to give a scientific certainty to the fantasy of authorship and autonomy that governs conventional masculinity. The film systematically presents but then undoes the foundations of such a fantasy, through its exposure of the illusions of the predictive certainty of genetic codes. One by one, each of the genetic predictions whose truths have justified the structures of *Gattaca*'s unjust society is undermined: the genetically guaranteed peaceful director of Gattaca turns to murder to defend his galactic vision; the doctor at Gattaca has an in-valid son, despite all the technological means at his disposal; the genetically perfect Jerome wins only the silver medal and consequently attempts suicide (leaving him in a wheelchair); and valid Anton cannot match in-valid Vincent's intellectual ingenuity or physical determination. These masculine subjects become the impostors against whom Vincent pits his willpower and intelligence: can he trick the director, can he become Jerome, can he beat Anton? In his fraudulent genetic identity, as someone who "postures as an autonomous and self-grounding subject" and yet can never be one, Vincent repeatedly performs his masculinity in relation to a series of masculine others who represent the threat of the "perpetual possibility of [his] own ungrounding" (Butler 1990, 45). The substance of his masculinity is tested through a series of challenges to the masculine impostors around him, who stand for the supposed genetic perfection of the valid sign.

Genetic perfection is the scientific equivalent of masculine singularity. As Marie-Luise Angerer (2000) has argued, from a Lacanian position, the fraudulence of masculinity lies in its imagined singularity. The imposture of masculinity is the imagining of the self as singular, as the one—the only one. In *Gattaca*, Vincent rehearses this fantasy. Can he become the author of his own desired identity? Can he prove he is the one? But Vincent's ambition to demonstrate the outstanding singularity of his masculinity, by pitting it against the superior claims of the valids, produces instead a relational identity, proving perfection in singularity to be an illusion.

Nowhere is the singularity of Vincent's heroic masculinity more clearly undone than in his collaboration with Jerome—for it is not one but two men who defy the laws of genetic determinism in *Gattaca*. Vincent's individual agency is dependent upon a collaborative, relational masculinity, for this is not a single imposture but duplicity, and, as the word suggests, this is a double vision (see the introduction for a discussion of the use of the DNA staircase here). This twofold agency requires an investigation of Vincent as not only an impostor, but also as an impersonator. Schwartz offers a definition of this distinction in relation to doubles: "imposture, the compulsive assumption of invented lives, and impersonation, the concerted assumption of another's public identity." He writes:

> Double agency, implying a singleminded performance of two opposed roles with silent devotion to a cause, is the impersonator's stock in trade . . . But impostors are unable to bear the burdens of double agency. Impersonation, not imposture, is at home with quiet deceit and may breed underground. Both may be impeccably costumed, yet in the final dressing down, impostors want attention and love, and we betray them; impersonators want our money, our secrets, our family line, and they betray us. (Schwartz 1996, 73)

If the masculine impostor is the one who appears to have achieved the singularity of autonomy and self-grounding, then the masculine impersonator is the one who recognizes the paradox of the need for the other in order to achieve this illusion. In queer debates, Butler has famously drawn on Esther Newton's work on drag queens in the United States to argue that "the structure of impersonation reveals one of the key fabricating mechanisms through which the social construction of gender takes place" (Butler 1990, 136–37).[18] Claiming that "drag fully subverts the distinction between inner and outer psychic space and effectively mocks both the expressive model of gender and the notion of a true gender identity" (ibid., 137), Butler refers to Newton's account of drag: "At its most complex [drag] is a double inversion that says 'appearance is an illusion.' Drag says . . . 'my "outside" appearance is feminine, but my essence "inside" [the body] is masculine.' At the same time it symbolizes the opposite inversion; 'my appearance "outside" [my body, my gender] is masculine but my essence "inside" [myself] is feminine'" (E. Newton 1972, 103). As Butler famously argued, drag makes visible the gaps between the anatomy of the performer, gender identity, and the gender being performed: "*In imitating gender, drag implicitly reveals the imitative structure of gender itself—as well as its contingency*" (Butler

1990, 137, emphasis in the original). Here repetition is crucial to thinking about the structure of impersonation. In her discussion of the problems of "writing as a lesbian," Butler suggests that "it is through the
repeated play of this sexuality that the 'I' is insistently reconstituted
as a lesbian 'I'; paradoxically, it is precisely the *repetition* of that play
that establishes as well the *instability* of the very category that it constitutes" (Butler 1991, 18, emphasis in the original). To the extent that
impersonation requires the repetition of duplicity, we might ask: does
it make visible the internal contradictions of identity that will lead to
its ultimate failure?[19] Extending feminist film theory through queer notions of impersonation, we might ask: what is specifically at stake when
the man is the site of repetition, of the ritualized reconstruction of masculine perfection as screen spectacle? In his impersonation of Jerome,
is Vincent caught in the same linguistic bind that will eventually ensure
his exposure and his failure?

To impersonate means "to invest with a supposed personality, to
embody," or "to represent in a personal or embodied form" (*OED*, definitions 1, 2). This suggests an artificiality that is the reverse of impersonation's authenticating intention (the desire to pass as someone else):
in the manufacture or production of authenticity, impersonation thus
both effects and undoes personhood. But how might we think about a
double impersonation, in which two bodies combine to produce a renewed version of genetic perfection? In impersonating Jerome, Vincent
invests himself with the bodily substance (the DNA sequence) and, by
implication, the "supposed personality" of another. It is their combined
labor that achieves a composite impersonation of masculine perfection:
Jerome produces his own bodily fragments and samples, and Vincent
reembodies them to produce the new Jerome.

If the masquerade functions as an imitation of a feminine identity
that is already a mask, what kind of masculinity is achieved through
the impersonation of genetic perfection? The concept of impersonation captures the new tension between image and identity invoked
by genetic engineering, which produces both original and copy. The
traditional binaries of identity formation (original and copy, authenticity and artifice, self and other) become redundant as biological reproduction is replaced by genetic selection, and recognition is based
on DNA profiling. If impersonation means to invest with personality or
bodily form, is genetic perfection not thus already an impersonation?

Genetic engineering invites us to intervene in future identities. In his impersonation of Jerome's supposedly unique genetic identity, Vincent confounds the truth claims of the predictions of techno-science and of the singularity of masculinity. His deception literalizes the question of the legibility of identity—of the language of genetics and the language of gender. The supposed transparency of genetic information is opened up to contested interpretations; the supposed authenticity of masculinity is exposed as a set of techniques that has become a marketable commodity. In the dynamic between Vincent and Jerome, masculinity shifts back and forth between them and becomes a transferable skill rather than an individual possession.

With the making of the new Jerome, there is a transfer of power and authority between the in-valid Vincent and the previously valid but now immobilized Jerome. The scenes of Vincent's early adult life show the impossibility of his becoming the self-grounding subject of masculinity: he is part of the anonymous workforce that polishes the shiny surfaces at Gattaca, in which the chosen ones will see themselves reflected back as the proper subjects they desire to be. But he cannot bear the repetitive work of cleaning (work typically associated with women, black people, and the working classes, but here crossing those divisions to be assigned to a new group of menial labourers, the genetically imperfect) and attempts to build up his own body in order to turn himself into a figure of proper masculine strength. Eventually, he recognizes he cannot do it alone; he needs a borrowed ladder to climb up the genetic hierarchy to achieve masculine agency.

Jerome, on the other hand, enters the film as a valid, marked through his disability as an in-valid, as I discussed at the beginning of this book. Although his disability excludes him from valid masculinity, he struggles to retain the superior status suggested by his air of upper-class English affectation, although such condescension is undercut by signs of his own physical and mental decline and decadence, indicative of a similar loss of social power and status. His disability marks the imposture of his previously valid status (winning only a silver medal and then attempting, but failing, to commit suicide). Despite his DNA, he cannot be the one. Jerome represents the desired masculinity of genetic selection, while simultaneously marking the fragility of its embodied capacity. Nothing seems to guarantee the successful embodiment of masculine perfection; its fragility is the only certainty. Valid masculinity is aligned with phallic potency in the scene of his painful attempt to ascend the spiral staircase—an echo of DNA's double helix—and answer the door to the detective, Anton, preventing him from discovering the deception.[20] Appropriately enough, Jerome's endeavor to retrieve his masculinity from its wounded or castrated symbolic state requires him to impersonate himself in order to reauthorize his masculinity. Accomplishing this almost impossible feat of passing as his previously able-bodied self, Jerome temporarily regains dignity, self-respect, and masculine integrity in the narrative. It is thus through impersonation that masculinity is repeatedly deauthorized and reauthorized throughout the film; its varying degrees are articulated in the two men's identities in highly relational terms.[21]

This complex relay of transferable identities brings with it a series of associations of kinship. Although Vincent and Jerome's relationship is nonbiological in the traditional genealogical sense, nevertheless it is all about shared biogenetic substances. Repeated close-up shots of bodily substances invoke a sense of shared embodiment through a nongenealogical kinship bond. The blood tie has been Western culture's mark of genealogy through kinship, but Vincent and Jerome reinvent kinship through the lens of the borrowed ladder. In this distinctly unconventional exchange of genetic material through prosthetic embodiment, the permanent and enduring ties of genealogy are replaced by a new relatedness.

Contrasting traditional notions of kinship ties as "unalterable biogenetic connections [that] accounted for the permanence of this very special sort of social relation" with new forms of relatedness, or "fictive kinship," Kath Weston explores the enduring loyalty and commitment of queer kinship systems (Weston 1991, 105, and 1998, 58). In *Gattaca*, Vincent's adoption of a new biogenetic identity occurs not only in the context of his rejection of his biological family but also, in the end, through a filial power struggle that he wins: in all the loyalty tests, "fictive kinship" wins out over genealogy. But perhaps "improvised kinship" is a better term for Vincent and Jerome's relationship.[22] As in all improvisations, this new form of kinship relies upon experimentation and risk, and also upon mutual trust and shared knowledge. While it is at first a pragmatic business deal in which the borrowed ladder is a commodity, the intimacy between the two characters soon goes beyond pure commercial necessity. Vincent and Jerome's improvisation produces a commitment typically reserved only for relatives or lovers: they share bodily substance; they willingly risk their lives for each other.

The improvised kinship tie between Vincent and Jerome is given symbolic permanence in the scene of Jerome's suicide, during which Vincent is reborn, and Jerome is released from life. Here their tie is transformed into a "forever" relationship, to use Weston's term: "Only a biological process (death), as opposed to a social process (rejection, neglect) is supposed to be capable of sundering 'blood' ties. In this reading death becomes the terminus that marks attainment of 'forever' in a relationship" (Weston 1998, 78). When the two finally separate (Jerome is about to commit suicide, and Vincent to go into space), Jerome gives Vincent a lock of his hair—an ironically romantic gesture in the wake of their genetic impersonation. With Jerome's death comes the confirmation of their permanent tie; in this romantic image of foreverness, the fantasy of a shared future is confirmed.

The connections between the forms of genetic impersonation and queer improvisation of kinship in *Gattaca* raise the question of how the heterosexual-homosexual distinction governs the place of sexuality in the culture of the copy. Throughout the film, the anxiety about detection is staged around what Eve Sedgwick has called "the relations of the closet—the relations of the known and the unknown, the explicit and the implicit" (Sedgwick 1994, 3). Vincent and Jerome's deception places them in a shared domestic space full of secrets, puts them at odds with society, and requires their utter loyalty to each other in the project of passing; their mutual trust is paramount to the success of their crime.[23] The relations of the closet that govern the staging of their conspiracy are reiterated in the homosexual connotations of the "perfect match" of Vincent and Jerome: the broker comments that "you two look so good together, I want to double my fee." The labor involved in the crime of genetic deception produces a physical intimacy between them that reinforces these associations. Vincent literally wears Jerome's body on his own. Jerome's gift to Vincent before his eventual suicide is to leave him samples of his bodily fluids—"enough for two lifetimes." Through what we might call the prosthetic intimacy of genetic impersonation, Vincent and Jerome share the most intricate knowledge of each other's bodies.

Even more explicitly, homosexual associations are articulated through Jerome's persona as the fading upper-class British retro dandy

19. Physical re-
semblance and the
expressive mirroring
of sly, secretive, and
proud looks connect
the arrogant Jerome
(Jude Law) (1) and the
enigmatic Irene (Uma
Thurman) (2) in *Gat-
taca* (Andrew Niccol,
1997).

and self-pitying, chain-smoking lush (see figure 19, 1). He embodies
many of the qualities of what Richard Dyer identifies as the "sad young
man" of 1950s and 1960s cinema: he is a figure of pity who is often
shown as melancholic, pathetic, and both "irredeemably sad and over-
whelmingly desirable" (Dyer 1993, 73). Squabbling on the telephone
with the hair-dye supplier about whether they have sent him "summer
wheat" instead of "honey dawn," he enacts the classic stereotype of the
"neurotic, hysterical, bitchy gay man" (ibid., 84). Vincent's and Jerome's
relationship reiterates many of the clichés of the conventional heterosexual marriage: Vincent is the husband who goes out to work and is
ambitious in the public sphere; Jerome stays at home and fusses over
domestic routines and cosmetics. He drinks more and more and be-

gins to act like a frustrated and jealous housewife, bitterly resenting Vincent's absences. Ultimately, the only meaning in Jerome's life is sharing Vincent's dream.

This homosocial intimacy, however, is disrupted by a third (female) term—the figure of Irene. This configuration is typical of what Sedgwick identifies as "a cultural system in which male-male desire [has become] widely intelligible primarily by being routed through triangular relations involving a woman" (Sedgwick 1994, 15). As the object of Vincent's romantic interest, Irene poses a threat to the exclusivity of the male bond between Vincent and Jerome.[24] But Irene also has an uncanny resemblance to Jerome: she shares his facial characteristics—high cheekbones, long jawline, steel-blue eyes, neat nose—and so mirrors Jerome's embodiment of sculpted white perfection (see figure 19, image 2). As such, Irene is the feminine counterpart or even the heterosexual equivalent of Jerome. Although Vincent's relationship with Irene is in some senses a traditional heterosexual romance, it might also be read through what Sedgwick calls the "projective fantasy" of "vicariated desire" (ibid., 156–57). The notion of "chains of vicariation" points to the interplay between identification and desire and between sameness and difference articulated in these triangular substitutions (ibid., 159). Thus, we might ask, with Sedgwick, "How is one to know whose desire it is that is . . . figured? By whom can it be so figured?" (ibid., 157).

This problem is amplified in a film about genetic impersonation: as an impersonation, Vincent is already a substitute for Jerome, and as a valid, Jerome is already artificially selected for his genes. Physical resemblance, sartorial matching, and rhymed gesture between Vincent and Irene is repeatedly shown in two shots that establish a flow of intimacy between them through gesture and expression (see figure 20). The genetic impersonations figure multiple chains of vicariation that disrupt the singularity of gender and sexuality and the authenticity of its embodied form. Vicarious desire for an impersonation presents an endless series of substitutions. Irene might be understood as a heterosexual object of desire, but one whose role is inextricable from the intimacy between Vincent and Jerome—an intimacy that is itself founded on a desire to become the other. In this context we might ask is Irene a substitute love object for Jerome, or is Vincent? When Jerome is feeling jealous, is he wishing he were Vincent or Irene? Is Vincent sexually appealing to Irene only in his impersonation of Jerome? Following the

20. The rhymed expressions of Vincent (Ethan Hawke) and Irene (Uma Thurman) as they watch a rocket launch in *Gattaca* (Andrew Niccol, 1997).

logic of this relay of queer substitutions through to its conclusion, we might ask if Jerome desires himself (in desiring to be Vincent, who desires Irene, who is a substitute for Jerome). Indeed, this may lead to the most disturbing fantasy about cloning: that we shall end up desiring ourselves.[25]

Gattaca presents a vision of the triumph of individual masculine desire that is repeatedly undercut by the doubly destabilizing and denaturalizing effects of these queer kinship improvisations and genetic impersonations. These substitutions are facilitated by impersonations which queer the singularity, intentionality, and directionality of desire. The end of the scene in which Jerome hauls himself up the staircase to impersonate himself illustrates this perfectly: "I think she likes us," he says to Vincent, including himself in Irene's plural object of desire. The composite fantasy of the new Jerome is the product of a homoerotic collaboration that is an impersonation of an impersonation: a vision of hypermasculine perfection as a cultural achievement in which genetic selection has already undermined the naturalness of identity and produced an artificially enhanced version of that idealized masculinity.

In the face of the undoing of masculine singularity by these prosthetic relationalities and vicarious substitutions, the individualist drive of liberal humanism that has haunted the whole film surfaces more fully in the penultimate scene as a vehicle for masculine heroism in the absence of its more foundational securities. The film presents us (perhaps somewhat ironically) with *Gattaca*'s vision of humanity as the sign of resolution, harmony, and hope. As Jerome prepares to commit suicide, Vincent's impersonation of him is revealed to have been no secret to Gattaca's doctor, who has tested his blood and urine since his first genetic interview at Gattaca. All the costly and devoted labor of scientifically precise disguise is rendered potentially redundant by one unconscious masculine gesture: "for future reference, right-handed men don't hold it with their left, it's just one of those things," he tells Vincent. Replacing the image of Vincent's in-valid identification card with Jerome's valid one, he confesses to Vincent about his in-valid son who would also like to work at Gattaca. The ultimate threat to expose Vincent's elaborate impersonation comes from a father with the power to prevent him from realizing his dream, who decides not to because he wants a sign of hope for his own in-valid son. He is a good father, who—unlike Vincent's—recognizes human potential in vulnerability and wishes to protect the imperfection that is human nature. Acting against the eugenic values of Gattaca, the doctor is a skeptical scientist who serves as a model of the humane gatekeeper; his humanity comes from a compassion born of proximity to genetic in-validity. As the film's tag line puts it: "There is no gene for the human spirit"—or, we might conclude, if there is, it is an in-valid one.

In contrast to the doctor's humanity, the corporation's vision of a panhuman future presented in the final scene in the rocket appears as artificial as its genetic engineering: the scene displays a markedly multicultural crew as reassurance against the associations of potentially racist eugenics of genetic determinism. As the camera pans around the inside of the spaceship, close-up shots of the many faces of the human race display the multiracialized diversity of a future eugenic humanism. Just as the image of a rocket launching into space as a symbol of Vincent's masculine achievement is a playful reiteration of a modern cliché of the gendering of technology, so the image of genetic technology producing a diverse global humanism leaves the audience with a familiar, and rather unconvincing, fantasy of unity through racialized diversity. This highly modern version of panhumanity is produced by technology (the vision of the blue planet from space in the 1960s), yet it also promises to resist technology (the potential evil of genetic determinism).[26] To be included in this fantasy of panhumanity, Vincent has first to be exposed, for "imposture is not imposture until its duplicity

is laid bare, and when impostors persist, treading in their own footsteps, they are not deranged but faithful to a lifelong project that oscil-
lates towards the spiritual" (Schwartz 1996, 71). If masculinity is sheer
imposture, and Vincent realizes his ambition in the very moment his
"duplicity is laid bare," the void left behind by such a disrobing might be
filled by the power of the human spirit.

If femininity as masquerade opened up the possibility of showing
the contradictions of the place of woman as image in a patriarchal sign
system—revealing in the end that there was, in fact, nothing behind the
mask but another mask—masculinity as impersonation points to the
dangerous illusory aspirations of singularity and perfection that govern
the drive for agency, self-grounding, and authorship on the other side
of the axis. But the display of the labor of artifice in the name of the
genetic impersonation of masculine perfection undermines any easy
attribution of gender, producing instead an ambiguity that plays across
the binary of sexual difference, queering the previous categories of feminist film theory. For the heterosexual-homosexual distinction takes on
a new significance in the culture of the copy, complicating the question
of how we know "whose desire is being figured" (Sedgwick 1994, 157).
The reproduction of sameness through sexual difference is no longer
so straightforward when the means for assuring its continuity are new
technologies of replication that threaten the authority of paternity, inheritance, and heterosexuality in the cultural imagination. Vicarious
sexual substitutions proliferate in the cloning cultures designed to imitate nature. It seems we should trust neither the cinematic nor the scientific evidence before us which promises perfection or predictability.
If the artifice of the image (femininity as the presentation of the desire
of the other) moves into the territory of the genetic imaginary, in which
technologies of cell replication provide the basis for fantasies for copying the self, then the battle over representation becomes, inevitably, a
battle over reproduction in both its traditional biological and cultural
senses.

From Bentham's "Panopticon" to Le Corbusier's architecture and Foucault's analysis of modern systems of surveillance, modernity's ideal of visual transparency found its most adequate expression in the concept of the big city.
STRATHAUSEN, "Uncanny Spaces"

Transparency has been a constant ideal in Western medicine, but that ideal has not remained static over the ages. Historically, this ideal connoted notions of rationality and scientific progress . . . In the twentieth century, the ideal of transparency has become associated primarily with notions of perfectibility or modifiability.
VAN DIJCK, *The Transparent Body*

[6] The Uncanny Architectures of Intimacy in *Code 46*

If "modernity's ideal of visual transparency found its most adequate expression in the concept of the big city," then postmodernity's flows of global mobility might find a reassuring anchor in the visual transparency of the human body promised by the blending of digital and genetic techniques. *Code 46* (Michael Winterbottom, 2003) offers a fusion of these two ideals of visual transparency: the regulation of the flow of people across the borders of a global megacity is melded with the screening of genetically engineered human bodies. In this exploration of a biotechnological future, the speedy movement of bodies across the borders of a zoned city is regulated through the genetic screening that renders the human body legible to the panoptic authorities of transnational corporations. The diegetic spaces of geographically recognizable locations (Shanghai, Seattle, and Jebel Ali) are connected through a network of screening technologies that tracks the movement of genetically engineered bodies across the globe.

Shaped by the lingering desire for control through visibility, which governed modernity's urban mappings, the genomic promise of a newly

legible bodily interiority appears finally to deliver that elusive fantasy
of ubiquitous vision. In other words, the notion that the genetic code
will render the human body transparent contains traces of the desires
motivating the mapping of modern urban architectures to regulate
subjects through optimum visibility. In *Code 46*, genetic engineering,
in the form of fetal cloning, is placed in a globalized world which, as
Peter Wollen puts it, has "moved us away from a time-based towards a
space-based culture, in which we can be constantly kept in touch with
simultaneous events from across the globe" (Wollen 2002, 214).

In a coalescence of these two ideals, the illusion of the transparent
geographies of modernity fuse with the desire for the transparent biologies of postmodernity: cartographies of urban space are blended
with the scientific mapping of the legible body through the idea of the
genetic code. In melding the control of inner and outer spaces through
the regulation of the geneticized body, *Code 46* presents a future in
which interior and exterior are no longer separable, and in which what
Anthony Vidler has called "the spatial relations of modernity" have been
transformed into one indivisible space (Vidler 2000, 76–77). Following
Vidler's treatment of space as "the complex exercise of projection and
introjection in the process of inventing a paradigm of representation,
an 'imago' of architecture . . . that reverberates with all the problematics of the subject's own condition" (ibid., 12), we might approach the
cityscapes in *Code 46* not as settings, but as projections of the modern subject's anxieties about the changing significance of the body's
interior.

When Elizabeth Grosz reverses the traditional formulations about
bodies and cities and asks if "architecture [can] inhabit us, as much as
we see ourselves inhabiting it," she offers us a way to conceptualize the
geneticized body in the cinema (Grosz 1995, 135). In this chapter, I address the question of how these two different kinds of mapping—geographical and biological—are given cinematic form. To what extent is
the gene given visual life through an architectural style that imagines a
spatialized form for this intangible dimension of the human body, which
is now present everywhere but visible nowhere? In what ways does
"modernity's ideal of visual transparency," expressed most adequately
"in the concept of the big city" (Strathausen 2003, 20), inform the ideal
of the "transparent body" (van Dijck 2005) in the postmodern "century
of the gene" (Keller 2000)? If the transparencies of internal and exter-

nal spatializations are connected through the continuities of urban and embodied architectural form, how might theories of the modern city be developed to offer an analysis of the cinematic mapping of the gene in the fluid spaces of globalized, cosmopolitan cultures? And finally, how are the genetically perfected bodies on the inside of the city in this film haunted by the less desirable bodies on the outside—that is, how is the mobile subject of global flow connected to those marked bodies restricted to the margins?

Since genomics is the science of genealogy, the geneticized body is always a spatialization of temporality: the passage of time across the generations is embodied in the concept of genetic inheritance.[1] In this film, time is spatialized and mapped onto the regulation of mobility across zoned territories, whose prohibitions function biologically and geographically. The particular use of postmodern, urban architecture in the film provides a spatialization of the recombinations associated with genetic engineering: the synchronic form of cartography maps out the diachronic temporalities of inheritance and relatedness. If, as James Donald has argued, "it is the attempt to render urban space transparent that produces the phantasmogoric city of modernity" (1995, 83), this film reworks the paradox of the modern city through the postmodern architectures of transnational mobility and genetic engineering. For Donald, the modern omnipotent fantasy of control through the *dieu voyeur* of the panoramic view of cities from above extends to the "dream of encompassing the diversity, randomness and dynamism of urban life in a rational blueprint, a neat collection of statistics, and a clear set of social norms"; he goes on to suggest that this is "an idealized perspective which embodies the enduring Enlightenment aspiration to render the city transparent" (ibid., 78). In *Code 46*, the attempt to render globalized spaces transparent through the regulation of mobility and flow, and transnational bodies transparent through genetic surveillance technologies, produces a journey not only across geographical space, but also into the labyrinthine, dreamlike, warped temporalities of the diegesis.

Code 46 is a stylish combination of noirish science fiction and romance film, which interweaves a cautionary fascination with technological innovation (genomic, computational, and visual) with a heterosexual love affair and classic noir motifs of mystery and deception through the mise-en-scène of urban night scenes, fast-tracked travel, and architectures of security and surveillance. The perfect site for narratives of deception and lost or mistaken identities, urban spaces have provided the location for all kinds of artificiality in the cinema, from the replicants of *Blade Runner* (Ridley Scott, 1982) to the robotic creations of *Artificial Intelligence* (Steven Spielberg, 2001) and the biovirtualities of *eXistenZ* (David Cronenberg, 1999).[2] The alienating cityscapes of science-fiction films play with the pleasures and dangers of anonymity, crowded spaces, and lost souls.[3] Just as *Blade Runner* places its replicants in a mise-en-scène of futuristic urban architecture, as Giuliana Bruno (1987, 66) has noted, so a number of more recent films about genetic engineering and cloning—most notably *The Fifth Element* (Luc Besson, 1997), *Gattaca* (Andrew Niccol, 1997), *Equilibrium* (Kurt Wimmer, 2002), *Teknolust* (Lynn Hershman-Leeson, 2002), *Casshern* (Kazuaki Kiriya, 2004), and *The Island* (Michael Bay, 2005)—use spatial motifs to produce a geneticized sense of future artificialities. As in *Blade Runner*, the urban location for this technologized future in *Code 46* is an Easternized one. Here, genetically engineered humans inhabit and embody the architectures of surveillance in a global city where speed and mobility are in tension with restrictions and corporate state control through a panoptic, technologized vision of the inside, as well as the outside, of the body.

This exploration of the convergence of genomic and surveillance technologies is played out through the story of William Geld (Tim Robbins), a happily married detective who arrives in a city called Shanghai from one named Seattle to investigate a woman, Maria Gonzalez (Samantha Morton), suspected of forgery of *papels* (papers needed for travel and insurance). William ends up falling in love with Maria, covering up her crime and ultimately fleeing with her illegally to *la fuera* (the outside), where the possibilities of escaping the genetic surveillance built into the urban architecture seem briefly to offer the promise of fulfillment of their romantic destiny. The narrative structure draws upon the genres of romance film (an unavailable man falls for a younger woman, but their love is prevented by an external barrier) and film noir (an enigmatic woman lures a man into a world of deception and crime, threatening the stability of his work, family, and moral values) but adds an Oedipal, science-fiction twist through the couple's genetic profiles (with a genetic profile almost identical to that of William's mother,

Maria becomes pregnant after her brief sexual encounter with him). When the transgression of code 46 (a law prohibiting sexual and reproductive relations between people too closely related genetically) is discovered by the corporate authorities, Maria is taken to a clinic, where they remove the fetus and her memory of the affair, implanting both a genetic virus to repel her from William in the future and a false, or prosthetic, memory of finger surgery to justify her visit to the clinic. When William discovers their code 46 transgression, he travels to find her at the clinic and take her home, but he never tells her of their genetic similarity and thus of their incestuous transgression of code 46. The couple then escapes to a guesthouse in *la fuera*, where they have sex for the second time in defiance of the genetically programmed state sanction (Maria invites William to override her repulsion of him with physical force). The narrative is structured around the question of whether heterosexual romantic love is more powerful than the joint forces of science and the global corporate state.

The initial sense of William's omnipotence, which is gradually eroded as his involvement with Maria continues, is established in the mise-en-scène of the arrival scenes. Opening with aerial shots of the end of a flight from Seattle to Shanghai, *Code 46* brings desert landscape and the big city into close proximity through a sense of the transparency of space afforded to the male protagonist (and the spectator) as he arrives in the global megacity of Shanghai, halted at checkpoints and borders where proof of his genetic identity permits his smooth entry into the zoned inner city. As Donald has written, "from [that] God's-eye view, the city presents a dehumanised geometry. People are . . . invisible, or . . . insignificant" (Donald 1995, 77). For William, the checkpoint authentications are as speedy and effortless as his movement across the globe. In this futuristic world of global genetic regulation, the legitimate mobility of citizens is policed through DNA screening which has become embedded in the routinized technologies of everyday security, whether through *papels* or instant genetic sampling. Shots of the desert sands— of the "zero degree of long and low-ness" of the desert, as Meagan Morris (1992, 11) puts it—contrast with those of the cityscape of Shanghai and its surrounding highways in William's smooth and rapid transition to the inner zone, prefiguring a series of more difficult journeys which punctuate the rest of the film. Each narrative event propels William into the next phase of mobility: there is hardly a mode of transportation

21. William Geld (Tim Robbins), the privileged global traveler, enters the inner zone of Shanghai in *Code 46* (Michael Winterbottom, 2003).

that he does not use in the film. The very few scenes of William's stasis seem to be mere markers along his incessant journeying. The on-location filming (in Shanghai, Dubai, and Jaipur), together with mobile framing and the use of a hand-held camera, produce a mode of spectatorship in line with William's journeying, even if not with his point of view (O'Riordan 2006).

In the opening sequence, William enjoys the ease of speed and movement through spaces almost devoid of people: the desert, the plane, the airport, and the road to the city. He inhabits them all with the privilege of calm and uninterrupted travel and solitary individuality. Unbridled by the competing needs of others, or the dull demands of domesticity or locatedness, the privileged white male moves freely across the globe, assisted by technology and the labor of others that function together to serve his needs. William seems to be the ultimate mobile subject who inhabits what Marc Augé (1995) calls the "non-places of supermodernity," spaces devoid of history, relationality, or identity, which are to be moved through but never fully inhabited: the airport, the five-star hotel, and the motorway. In contrast to the lot of would-be migrant laborer stopped at the threshold to the inner zone of the city, William's speed and mobility belong to the universal register of "global cultural flow," not to the local restrictions of time or place (see Appadurai 1996). The non-white migrants who crowd around William's taxi embody the other side of postmodernity's fantasies of global mobility (see figure 21): those fixed exclusions whose immobility confirms the flow of the privileged few (see Ahmed 2000). Like Augé's nonplaces, where "transit points and temporary abodes are proliferating under luxurious and inhuman conditions (hotel chains and squats, holiday clubs and refugee camps, shantytowns threatened with demolition or doomed to festering longevity)," the anonymous sites of both William's transnational movements and the bleak borders of access to the inner zone share a sense of a world "surrendered . . . to the fleeting, the temporary and the ephemeral" (Augé 1995, 78).

In the early scenes of the film, William moves fluidly across a series of spaces governed by an aesthetic of repetitive forms and sequential structures: in the hotel, for example (see figure 22), and in The Sphinx, the corporation where Maria works. The even, genetic geometries of the mise-en-scène are matched by the cyclical, mesmerizing repetitions of The Free Association's electronic, minimalist musical score. Throughout the transition from desert to big city, from air to road transport, bird's-eye shots of the changing terrain produce a familiar sense of a God's-eye visual transparency, mapping out a clear spatial grid of progressive technological development—from empty desert to scattered small holdings, villages, towns, and finally city (see figure 23, image 1). Point-of-view shots of this changing landscape from the plane combine with the repeated use of long shots of a virtually solitary traveler, surrounded by empty space, to produce a sense of expansive spatial privilege. Aerial shots of William's arrival at the airport, on the highway, and in the hotel emphasize the absence of queues, crowds, traffic jams—almost of other people (see figure 23, image 2). Here, the mutuality of transparency and mobility offer the omnipotent pleasure of speed and ease William's arrival in Shanghai.

This fluid global mobility is enacted through the fast-moving, multi-lingual dialogue. The lingua franca of the film is an ersatz language derived from numerous sources, including English, Spanish, French, and Arabic. William and Maria's shared knowledge of this recombinant language (despite the geographical distance of their habitats) immediately unites them within the proximity of shared global citizenship. The shifting language moves seamlessly from one national register to another, denoting a similarly blended subject whose ease and rapidity of movement is indicative of this future form of biocitizenship: the transcendental subject whose familiarity with international codes apparently allows an infinite cultural, as well as geographical, mobility.

Although these early scenes fuse the visual sense of spatial transparency with William's smooth mobility across borders and security checkpoints, the consequent omnipotence afforded him is placed in

22. Top: The architectural style of William's hotel in
Shanghai emphasizes repetition and symmetry in *Code 46*
(Michael Winterbottom, 2003).

23. Aerial shots of rural (1) and urban (2) landscapes give
William's journey into Shanghai a sense of his mobility
across vast unpopulated spaces in *Code 46* (Michael
Winterbottom, 2003).

tension with an anonymous female voice-over, which undermines the authority of William's screen presence. The woman's direct address to William establishes a romantic intimacy between them: "I think about the day we met—I suppose you would have arrived *par avion*. Maybe you were first to get to security. You didn't intend to stay. You only had 24-hour cover, so luggage *a mano*, and they probably had a driver meet you so you didn't need to find a *coche*. You'd never been to Shanghai before." This voice-over is given diegetic agency as it narrates William's activities for the spectator in the time immediately preceding his first meeting with Maria. Shots of William arriving and in his hotel room obediently follow what the voice describes: "You arrive in the morning so the streets are deserted. Did you call anyone, did you have a *cerveza*? . . . Sometimes I imagine you watching the news, sometimes playing a game, sometimes sleeping, preparing yourself for the night ahead. The thing I can't imagine is that we hadn't met, we hadn't even heard of each other, that you thought it was going to be just another night." Her anticipatory retrospection[4] articulates a voyeuristic nostalgia for a past which has yet to connect with her own, conveying a kind of postmodern belatedness in the romantic fantasy of intimacy, predating their future encounter. Together with the slow, repetitive, electronic music, this wistful voice-over combines flashback, nostalgic fantasy, and romantic daydreaming. As in Freud's discussion of the child who imagines himself to have been present at his own conception,[5] this voice-over gives Maria a presence in a past already lost to her. This retrospective omnipotence functions as an aural equivalent to the protagonist's God's-eye view of the changing landscape from the plane and as direct competition for authorial control, raising the question of whose story this is, William's or Maria's.

This tension is reinforced through a familiarly gendered dichotomy: William's "panoramic vision" as he enters Shanghai is in contrast with what Michel de Certeau calls the "oneiric figuration" of the city as a phantasmagoria (de Certeau 1984), which is aligned with what turns out to be Maria's point of view (de Certeau, quoted in Donald 1995, 81). While William is the privileged outsider who views the city from above, Maria embodies the city in the sense that she inhabits its underground capillaries and hears its whispered secrets. As Donald has suggested, "the great figure [in] this confrontation between the transparent, readable city and the obscure metropolitan labyrinth . . . is that of the detective" (ibid., 79). He promises a solution to what de Certeau (1984) calls the confrontation between the panoramic desire to encompass "the diversity, randomness and dynamism in urban life in a purified, hygienic space . . . of benign surveillance and spatial penetration" and the mythic notion of the city as "a labyrinthine," dream-like figuration (Donald 1995, 78). William enters Shanghai in *Code 46* as the genomic extension of the modern desire to make the city governable through transparency. Like the human equivalent of the genetic screening technologies which make the body legible in the film, William can read people's minds. Appearing at first intuitive, but then revealed as a genetically engineered viral implant, William's empathy enables him to detect crime on behalf of his corporation, Westernworlds, through genetically produced feelings for another (empathy means "to share a feeling with someone"). As amplified bioempathy, which is just another aspect of global genetic regulation, William's virus represents the ultimate fantasy of transparency through immediacy—a rebiologization of affect through genetic engineering.

The choice of Shanghai as the location for a story about the regulation of highly dispersed populations through rigid, centralized modes of genetic and surveillance technologies is not insignificant. If "the geography of globalization contains dynamics of both dispersal and centralization," then Shanghai demonstrates precisely the problems of national and state regulation in the context of the fluidity and mobility demanded by the expansion of global capitalism (Sassen 2002, 3). As Saskia Sassen argues about globalization generally, the tensions between the demands of cross-border "flows of capital, labor, goods, raw materials and travelers" and national state regulation have taken on particular intensity in "the formation of new geographies of centrality in which cities are key articulators" in global circuits (ibid., 1–2). One of the new "Asian-Pacific megacities" of the global age (Gu and Tang 2002, 273), Shanghai condenses the tensions between fluidity and mobility on the one hand, and restriction and state control on the other hand. While Shanghai regained its central status in the Chinese economy when Pudong was designated as a special development zone in the early 1990s, placing it at the forefront of the country's economic reform, Shanghai continues to be subject to the heavy restrictions of communist state regulation (ibid.). The city has made major efforts to internationalize and modern-

ize itself, which have had an impact upon its urban spatial and func-
tional structures. A "gateway to the future," it has been subject to rapid
transformation in the race to establish Asian-Pacific "business hubs for
the regional headquarters of multi-nationals, and for the specialized
corporate services" which surround them (ibid., 273). Shanghai has
come to be seen as a non-Western, cosmopolitan city, whose history
combines associations of modern authoritarian state surveillance with
the postmodern architectural styles of multinational buildings designed
to make their mark on this rapidly changing city.

This tension between rigid state regulation and the free flow of goods
and people that has come to mark Shanghai's reputation and history
echoes the paradoxical promise of genetic engineering and screening
technologies in *Code 46*: on the one hand, genetically engineered bodies appear easier to regulate (tracking through DNA screening); on the
other hand, the legibility of geneticized bodies poses new problems for
global corporate control of Westernworlds and The Sphinx. Shots of
Shanghai's cityscape in *Code 46* offer a visual analogy for the mixed
biologies of genetic engineering where the disembedding and reembedding of biological matter threaten the foundations of life itself: the
echoes between inner and outer mappings are built into the mise-en-
scène as the recombinant moves of genetic engineering are repeatedly
connected to the blended styles of the city's design.[6] Both these recombinant forms undermine traditional notions of singularity and originality,
drawing attention to their imitative structural analogies and posing the
question of how to read the relationship between surface and depth,
between appearance and substance in each case. These analogies can
be extended further into the film's internal tensions between its devotion to traditional modes of verisimilitude (on-location shooting and
the use of 35-mm film) and its multiple citations of the histories of the
genres of noir, science fiction, and romance: *Out of the Past* (Jacques
Tourneur, 1947) meets *Brief Encounter* (David Lean, 1945) meets *Blade
Runner* and *Gattaca*.[7]

Code 46 contrasts the ancient with the (post)modern, the past, and
the future, through very particular visions of the East. The simultaneous presence of competing modes of inhabiting the inner and outer
zones in the film operates through the coalescence of different notions
of the East as non-Western. As in *Blade Runner*, the East (and the so-
called Orient in particular) here provides the imaginative location for a

tale of rapid technological development. As Sharalyn Orbaugh (2005) points out, this techno-orientalism is part of a long tradition in Western popular culture, through which the East is depicted as both envied and admired by the West.[8] Setting a cloning narrative in Shanghai in this way places fears about genetic similarity and physical resemblance within the history of Western Orientalism, which has typically cast Oriental people as indistinguishable, inscrutable, and enigmatic—like clones, they all look the same (see Said 1979). Alongside the postmodern urban architectures of science-fiction films are the low-tech desert of exile and the romantic free port of Jebel Ali, where escape from surveillance seems briefly possible for the eloping couple on the run. In *Code 46*, fetal cloning interferes with the universal laws of nature (the incest taboo) and requires cultural regulation through legal codification, stretching the temporal associations of the East with the ancient traditions of the past and the future of technical innovation simultaneously.

The desire for visual transparency in the urban architectures of *Code 46* is related to the problem of misrecognition in the face of the deceitful surface appearances of reiterative and recombinant forms. The question of the visual transparency of genetically engineered bodies presents the problem of recognizability. The genetic law, code 46, instantiates a series of prohibitions—purportedly to protect people from the incestuous dangers of their own desires—precisely because genetic kinship (with its potential for excessive proximity) cannot be made visible to the naked eye, on the surface of the body. As the film begins, this text appears across the background of desert landscapes like an ancient law:

> Article 1.
>
> Any human being who shares the same nuclear gene set as another human being is deemed to be genetically identical. The relation of one are the relations of all. Due to IVF, DI embryo splitting and cloning techniques, it is necessary to prevent any accidental or deliberate genetically incestuous reproduction.

As this text from the opening shots of the film explains, code 46 is a law that requires the screening of all fetuses of unplanned pregnancies and of all prospective parents; it prohibits pregnancy in cases of shared genetic identity of more than 25 percent, and requires termination of all fetuses resulting inadvertently from such couplings. The law also criminalizes those who break it and requires them to submit to medical treatment to prevent them from violating it in the future. The problem of making identity legible, by making visible the hidden genetic truths of kinship, involves a whole set of interrelated questions of sameness and difference, some requiring specialist or expert literacy techniques. Since sameness and difference can no longer be read off the surface of the body, but rather threaten to undo us from the hidden depths of genetic sequencing, people have much to fear in *Code 46*. The geneticized body is rendered tangible through spatialization: for example, William's genetically engineered empathy virus (which he uses to gain privileges by exercising power over people) only works in the urban spaces of the inner zone. The problem the film poses might be framed thus: if the much proclaimed visibility of sexual difference and the socially desirable non-Oedipal structure of exogamous heterosexual reproduction are to be replaced by the artificial kinship structures of genetic engineering and fetal cloning, how are we to recognize the differences that matter in that determining moment of desire?

Vidler offers an important challenge to Foucault's interpretation of Enlightenment space, which he argues is limited by Foucault's "insistence on the operation of power through *transparency*, the panoptic principle, [that] resists exploration of the extent to which the pairing of transparency and obscurity is essential for power to operate" (Vidler 1992, 172, emphasis added). Instead, Vidler proposes that it is in "the intimate associations of the two, *their uncanny ability to slip from one to the other*, that the sublime as instrument of fear retains its hold—in that ambiguity that stages the presence of death in life, dark spaces in bright spaces" (ibid., emphasis added). Drawing on Vidler's claims that the uncanny is "literally built into both the metropolis and the cinematic apparatus," Carsten Strathausen argues that the uncanny captures "modernity's oscillation between exposure and repression, between location and displacement" (Strathausen 2003, 15–17). Freud's concept of the uncanny starts with the "disquieting slippage between a place where we should feel at home"—the familiar—and "the sense that it is, at some level, definitively unhomely" or "*unheimlich*" (Donald 1995, 81). Vidler extends the problem of "anxiety and the paranoid subject of modernity beyond the question of the domestic uncanny (literally the 'unhomely')" to consider

the idea of phobic space and its design corollary warped space, understood as a more general phenomenon touching the entirety of public territories—the landscapes of fear and the topographies of despair created
as a result of modern technological and capitalist development, from Metropolis to Megalopolis. (Vidler 2000, 2)

Vidler's conceptualization of "warped space" provides an entry point
for a consideration of the uncanny spatial dynamic within the cinematic
language of the gene. Building on modern fears of the double lurking in
the shadows of urban spaces and on a long tradition in science-fiction
cinema, current films about genetic engineering and cloning focus on
the loss of distinctions between life and death, between self and other,
and between inside and outside (Telotte 1995 and 2001). For Strathausen, the uncanny is present in both the cinema and the modern city,
since each is premised on something presumed dead being "brought
back to life" and beginning "to haunt the living" (Strathausen 2003,
17). But, he argues, it is not only darkness which haunts the urban
spaces of modernity (with all the colonialist connotations which inform
its racialized narrative of Western progress), it is also multiplicity (ibid.,
18). Multiplicity, he suggests, threatens the singularity and individuality
which lie at the heart of modern aspirations to subjecthood.

With the advent of cloning techniques, multiplicity has now come
to haunt not just the subject, but also previously sacred notions of the
singularity of embodiment and of linear genealogy through the sanctity
of heterosexuality and reproduction. As a "technology of haunting,"[9]
genetic engineering and cloning enact a godlike intervention into the
teleologies of life: resurrecting and duplicating life (*Alien: Resurrection*
[Jean-Pierre Jeunet, 1997] and *Godsend* [Nick Hamm, 2004]); predicting and controlling future generations (*Gattaca* and *The Island*); preventing disease and holding the body at the threshold of death (*Lorenzo's Oil* [George Miller, 1992]); and bringing the past and future into the
present (*Jurassic Park* [Steven Spielberg, 1993]). The figure of the human clone in the cinema (*The 6th Day* [Roger Spottiswoode, 2000] and
Multiplicity [Harold Ramis, 1996]) embodies the uncanny possibility of
human doubling, drawing on longer traditions of the fascination with
the double that are now imagined through genetic engineering.[10]

If "the virtual fears of late modernity . . . bear at least a family resemblance to the old phobias of modernity," as Vidler (2000, 1–2) suggests, then current anxieties about cloning extend the uncanny into the

imagined possibilities of multiplicity in the new realm of biosecurity for reproductive and genetic engineering. The transparency of embodied identity promised by genetic surveillance seems at first to solve the problem of how to regulate highly mobile populations. But the transformation of the body into information produces endless possibilities for imitation and impersonation. Thus, we might conclude that just as "the Enlightenment dream of rational and transparent space was troubled from the outset," as Vidler (ibid., 3) argues, "by the realisation that space as such was posited on the basis of an aesthetics of uncertainty and movement and a psychology of anxiety," so the dream of the body made transparent through genomics is also haunted by the inevitable failure of technology to secure identities and boundaries, whether around the subject or around the spaces of urban and national territories.

Code 46 brings what Vidler calls the "anxious visions of the modern subject" into the realm of genetic engineering, where the contemporary subject's body is not only "caught in a spatial system beyond its control" (Vidler 2000, 1), but regulated by a security system operating at an unprecedented speed across global networks. In the rest of this chapter, I explore the problem of the transparency of the geneticized body in Code 46 in relation to claims about modernity's failures to exorcise the uncanny.

The current concern with the transparency of the genetically engineered body extends Vidler's reading of uncanny urban spaces. Like the medical-imaging technologies in van Dijck's study of the cultural ideal of the transparent body, DNA screening promises a vision of clarity and control that is undeliverable precisely because of the uncertainties produced by the technologies themselves. As van Dijck argues:

> Imaging technologies claim to make the body transparent, yet their ubiquitous use renders the interior body more technologically complex. The more we see through various camera lenses, the more complicated the visual information becomes . . . the mediated body is everything but transparent; it is precisely this complexity and stratification that makes it a contested cultural object. (Van Dijck 2005, 4)

The contradictory notion of that which is transparent as "having the property of transmitting light, so as to render bodies lying beyond completely visible" and as "easily seen through . . . or detected" (OED, definitions 1a, 2b) is central to the shifting relations of vision, knowledge, and the body in Code 46. Undermining conventional associations of light with transparency, the typical activities of day and night are reversed because of the carcinogenic threat of daylight. As they walk together in the deserted city, William and Maria hide under cover from the dangerous rays even when the sun is low in the sky (see figure 24). In the glaring daytime heat, as they rush through the dangerous, hyperilluminated, urban spaces devoid of people, fleeting and sun-dazzled shots deny the spectator any sense of visual clarity. Similarly, Shanghai's urban locations both offer and refuse knowledge through vision: in scenes of the city in daylight, shots of the glass surfaces of skyscrapers reflect back a mirror image of the outside rather than offering a view of the inside; and in interior locations, such as William's hotel room, windows both offer a view of the sunset skyline, with the city's famous Pearl Tower outside, and double as screens for news and entertainment for the guest inside. Throughout the film, the problem of how to make something transparent (meaning to appear across) is built into the mise-en-scène. As walls turn out to be mirrors and windows transform into television screens, the multiple appearances of transparent surfaces reinforce the increasing ambiguity of the inner truth of the couple's genetic kinship. Since transparency promises that something will both reveal itself and reveal something else beyond it, the potential for misrecognition is embedded in the very desire for singularity at stake in visual clarity.

Like *Species* (Roger Donaldson, 1995) and *Gattaca* (see chapters 3 and 5, respectively), *Code 46* is governed by the desire to see through (both literally and figuratively), yet punctuated by the repeated rehearsal of the impossibility of any certainty that vision might afford. In both *Gattaca* and *Code 46*, deceptive reflections question not only the status of the genetically engineered body but also the authenticity of the image, bringing together the suspect artifice of both surface appearances. As Jay David Bolter and Diane Gromala suggest in a very different context, the philosophy of transparency has a history. "In painting [it] meant creating an image that fools us—makes us think we are looking at a physical world" (Bolter and Gromala 2003, 36–37). The hallmark of modernist design was "to ensure that [it] would be immediately readable." If they are right to claim that the "myth of transparency" is about origins, then its centrality to this renewed desire to make the body legible through genetic engineering and screening techniques extends questions of vision into the domain of corporeal control (ibid., 48).

24. The lovers, William Geld (Tim Robbins) and Maria Gonzalez (Samantha Morton), seem to have Shanghai all to themselves as they cross the urban landscape in *Code 46* (Michael Winterbottom, 2003).

The illusions of visual and corporeal transparency converge in the mise-en-scène of corporate techno-science of *Code 46*. Since William is an agent of the multinational fraud detection agency in Seattle, Westernworlds, whose task is to solve the crime of *papels* forgery at Maria's workplace, The Sphinx, his capacity to see through deception is crucial. The promise of transparency in the design and layout of The Sphinx is undone by the tension between uniformity, regulation, and order, on the one hand, and the secret routines of forgery, on the other hand. A mise-en-scène of stylish modernist minimalism produces a sense of transparency through repetition, sequence, and uniformity. Reminiscent of the workplace interiors in *Gattaca*, The Sphinx has rows of glass-fronted workbenches with no privacy or distinction; an open-plan layout; uncluttered surfaces; a décor of clear blue, gray, and white; and controlled artificial interior lighting (the absence of daylight). The high-angle, long shots of the laboratory maximize its scope and efficient observation and surveillance, producing a sense of omnivision for both the spectator and the authorities (see figure 25). The appearance and behavior of the multiracial workforce reiterates the principle of uniformity across diversity. But despite this veneer of order and obedience, the apparently transparent spaces occlude criminal activities, as Maria and her collaborators continue to escape the controlling gaze of global corporate regulation. The sense of calm, predictable order in this workplace is achieved through a geneticized aesthetic of symmetry,

25. The rows of identical workstations at The Sphinx, where Maria works
in *Code 46* (Michael Winterbottom, 2003), are laid out in open-plan
style to permit maximum surveillance.

balance, and repetition, behind which deception proceeds undetected,
facilitating the continued mobility of illegitimate subjects across the
forbidden borders of the zoned city outside.

The modes of regulation in *Code 46* are in many ways less like the
disciplinary surveillance of modernity challenged by Foucault, and
more like what Gilles Deleuze has described as "societies of control" in
which power mutates in the "ultrarapid forms of free-floating control
that replaced the old disciplines in the time frame of a closed system"
(Deleuze 1992, 4). Unlike the previous disciplinary "enclosures" that
were "molds, distinct castings," these new "controls are a *modulation*,
like a self-deforming cast that will continuously change from one moment to the other, or like a sieve whose mesh will transmute from point
to point" (ibid., emphasis in original). Whereas the "disciplinary man
was a discontinuous producer of energy . . . the man of control is undulatory, in orbit, in a continuous network" (ibid., 5–6). The sense of continuous tracking regardless of location distinguishes these new forms of
power from their disciplinary predecessors through "the conception of
a control mechanism, giving the position of any element within an open
environment at any given instant (whether animal in a reserve or human
in a corporation, as with an electronic collar)" (ibid., 7). In this new regime of universal modulation through screening and information technologies, the rules may change, the permits may expire, and the subject's
status might shift, as control "is short-term and [has] rapid rates of turn-
over, but [is] also continuous and without limit" (ibid., 6). *Code 46* begins

with panoptic fantasies of the cosmopolitan corporate traveler survey-
ing the landscape below from the comfort of his plane, but these soon
unravel as he is drawn into the uncertainties of the shifting ground of
universal modulation that undermines his potency.

The urban spaces of *Code 46* also resemble elements of Felix Guattari's imagined city (described by Deleuze), in which "one would be able
to leave one's apartment, one's street, one's neighbourhood, thanks to
one's (individual) electronic card that raises a given barrier; but the card
could just as easily be rejected on a given day or between certain hours;
what counts is not the barrier but the computer that tracks each person's position—licit or illicit—and effects a universal modulation" (ibid.,
7). The unpredictable flow of genomic information through networks of
global mobility in *Code 46* repeatedly eludes the tracking techniques
of corporate surveillance. Instead, new forms of ubiquitous and fluid
monitoring accompany the subject who traverses the inner and outer
zones of transnational biosecurity. In this control society, invisible and
unpredictable changes to William's and Maria's privileges of mobility
(his empathy virus, her thumbprint, his *papels*, her forgeries, and his
and her prosthetic memories) keep shifting the ground rules of the surveillance systems. If, as Deleuze argues, a control society is a "universal system of deformation" and of "limitless postponements," then the
impermanence and mutability of this new order of control produces a
constant sense of instability: yesterday's privileged knowledge becomes
today's liability; today's source of stability becomes tomorrow's threat
of infiltration. In this "undulatory" and "discontinuous" set of distribu-
tions, fetal cloning and genetic screening constitute the body as just
such an unpredictable and shifting site of insecurity (ibid., 5).

The replacement of the Fordist factory by the multinational corporation is integral to the shift from discipline to control (ibid., 6). In this
context, Deleuze suggests that "what is important is no longer either
a signature or a number, but a *code*: the code is a *password*" (ibid.,
5, emphasis in the original). *Code 46* blends all three meanings of the
word: "code" refers to the legal foundation for the inner zone, to a
healthy kinship system of exogamy (conforming to the prohibition of
incest), and to the genetic information contained in DNA. In this world
of fetal cloning, code 46 is implemented by corporations that govern by
reading people's genetic codes as the new passwords for the regulation
of kinship, sexuality, and reproduction. Managed by a global network

of "corporations with souls" ("The Sphinx knows best"), genetic codes become the passwords through which mobility is regulated: William's movements are tracked as he moves across the globe; as screens appear and disappear, technologies mutate, and information flows through the invisible architectures of genomic surveillance.

In combining anxieties about the biological legibility of authentic identity, kinship, and relatedness in a world of fetal cloning with contemporary fears concerning geographical security, border control, and the mobility of migrant populations, *Code 46* presents a vision of multiculturalism that is haunted by both the legacies of a colonial past and the potential threat of fluidity in a globalized future. As Strathausen, drawing on Foucault (1980), writes, the Enlightenment and cultural modernism shared a "preoccupation with an elucidating visibility rendering the dark spaces of the outside utterly transparent" (Strathausen 2003, 15). Such fantasies of control of urban life and space articulate a colonial anxiety which attaches to darkness a set of associations with the threatening unknown, and to light the reassuring promise of knowledge through clarity of vision. The desire for transparency works through a racialized register that promises to defend the modern subject against the multiple hauntings of the postcolonial era, in which, as Sneja Gunew suggests, the projects of nationhood and multiculturalism are haunted by their colonial legacies and by the ghosts of the "attempted genocides and associated atrocities of colonialism" (Gunew 2004, 129). The control society of *Code 46* attempts to regulate an accelerated flow of populations across geographical spaces of a postcolonial world that is terrified of the invisible other within, as it prohibits entry to the undesirables without.[11] The stability of the inner zone is threatened by the return of the repressed, in the form of the displaced, dispersed, and migrant laborers.

Postnational and postpolitical, society in *Code 46* has replaced the state with global corporations that regulate human relations through genetic policing. In this fantasy of ubiquitous vision through dispersed networks, power is exercised at the most local, and supposedly the most conclusive, level: that of the gene. A microcosmic articulation of regulation, the gene becomes cellular information which is subject to surveillance. If the flow of people across borders in a globalized future proves to be a danger to those who nonetheless desire it, then the certainty of genetic identity becomes the logical form of security.

Patterns of complex dispersal and mobility in the postcolonial era are regulated by "ultrarapid forms of free-floating control" (Deleuze 1992, 3), which rely on the information flows of globalized networks. In an attempt to prohibit illegal entry, borders are policed through security and surveillance technologies: DNA fingerprinting, iris scanning, and computerized passport control. To use Deleuze's words again, as the "mesh of the sieve" mutates to regulate undesirable outsiders, genomic screening produces shifting categories of exclusion: the migrant, the criminal, and the eloping couple.

Genetically engineered bodies in this film are inscribed within a tension between the universal mobility of the blended subject of panhumanity and the threat of those designated genetically undesirable, who clamor at the highly regulated borders of the inner zone.[12] In contrast to the restricted movement and limited visions of those on the ground and at the borders, the blended panhuman subject embodies the fantasy of universal presence through swift mobility in the rapidly globalizing flows of postmodernity. In its blended formation, the panhuman subject is the sophisticated embodiment of a multicultural cosmopolitanism, of an ease with East as well as West, enacting a desire to grasp the totality of the world, and satisfying a fantasy of omnipotent vision necessarily always beyond the subject's psychic grasp. This illusory sense of global mastery through macrovision here might be read as the counterpoint to the mysteries of microscopic life at the genetic level (see Roof 2007).

As the inner zone of the future, the Shanghai of *Code 46* has a decidedly multicultural feel: members of the ethnically diverse workforce at The Sphinx list their various national and regional origins for William like a condensation of a globalized genealogy. This sense of diversity extends into the scenes of William and Maria's growing romantic attachment in the crowded night market and the metro, where the multicultural population inhabits the cosmopolitan spaces of Shanghai with ease in what Anne-Marie Fortier (2008, 68) calls "inter-ethnic propinquity."[13] But the sense of global scope and instant presence for the panhuman subject in *Code 46* is troubled by the potential threat to its potency and security. This vision of the future (like the inverse of the colonial project of racial segregation) is of a racially mixed and nationally diverse population that inhabits the privileged inner zone, while the outside is populated by more racially uniform groupings of nonwhite, non-Western people who represent the technologically unmodernized.

The desire for absolute control of sexuality and reproduction is thus mapped onto the regulation of mobility through surveillance in a zoned world, where the threat of the outsider is given geographical location and physical presence. Just as the postcolonial project of the multicultural nation is haunted by the past atrocities of colonial modernity, this vision of the genetically suitable, panhuman subject of corporate global culture is surrounded by the exclusions necessary for this new system of privilege. While genetic screening promises the security of a new biological transparency, the ghosts of the undesirables hover at the margins. These people are the undead, the nonliving who inhabit those "abject zones within sociality" (Butler 1993, 243): "How do people live out here?" asks William when he first arrives. "It's not living," replies his taxi driver. Here, the panhumans require the restricted mobility of the not quite fully humans to designate their genetic and other invisible biological superiority.

But the narrative trajectory that moves from privileged inner zone to spaces on the outside in *Code 46* does not unambiguously endorse the former over the latter; neither can the film be read through the rigid separation between the desirable inside and the undesirable outside, as the taxi driver's comment to William might suggest. While those who clamor for entry at the borders of Shanghai represent the desperation of the excluded, the extended cooperation of the community around the guesthouse to which the couple flee offers an alternative set of values through its slower pace and calm appreciation of the earthy aesthetic of terra-cotta colors, hand-woven fabrics, and low technology. Moreover, Maria is never fully at home on the inside. Even if her own encounter with the outside turns out to be a brutal one, her criminal forgeries of *papels* allow others to live out her dream for her and repeatedly activate a nostalgic longing for it associated with her father (who was banished to the outside by the authorities).[14] The outside thus holds both the threat and the promise of another way of life: its ghostly traces surface in the urban spaces of Shanghai even as the city's security systems apparently keep them safely banished.

Both geographical and biological spaces are constituted through fears about the regulation of boundaries between the inner and the outer in the absence of stable visible evidence. The entry of the detective into the urban spaces of science-fiction films "stages the city as enigma: a dangerous but fascinating network of often subterranean

relationships in need of decipherment" (Donald 1995, 79). Like his generic predecessors, such as Deckard (Harrison Ford) in *Blade Runner*,
William has the job of turning the city into a "purified, hygienic space"
(Donald 1995, 78), purging it of what de Certeau (1984, 94) refers to as
"its physical, mental and political pollutions" by solving a mystery or
crime. In this sense, we might echo Larissa Lai's concern that in science
fiction generally, the white "detective figure's ability to solve the crime
is directly connected to his racial purity" (Lai 2004, 27). As the white
detective, William at first appears to be the embodiment of how the disciplinary modes of colonial modernity might mutate into the informational reach of the global network of control societies. The only white
male character of any significance in the film, William stands alone as
a symbol of the ubiquity of corporate capitalism and of the certainty
of the new transparencies of mind and body in the age of genetic scrutiny. In an early scene of William's fraud detection, when he interviews
employees at The Sphinx, his empathy virus represents the ultimate
fantasy of a new ubiquitous vision: he remains the white clairvoyant
anchor (the one who knows and can see through people) in contrast
to the multicultural diversity of the suspects' geographical and ethnic
origins, shown in a rapidly edited, jolting sequence of close-up interview footage that resembles a sped-up series of police interrogations or
documentary interviews.

   At the opening of the film, William displays the potency of the white
detective of science-fiction films; however, he is quickly robbed of this
power through a classic noirish route: his desire for a woman—Maria,
the highly enigmatic, if rather boyish, femme fatale. Once William's
own body has betrayed the laws of his corporate world (his lie to both
corporations about who was forging papels, together with his transgression of code 46 by having sex with Maria), he begins to move out of
the inner zones and into the marginal spaces of la fuera (the abortion
clinic, the Freeport, and the desert) where his empathy virus no longer
functions to deliver his privileged access and mobility. There he stands
alone, stripped of his white, phallic potency. To make the point in the
most obvious psychoanalytic terms, the genetically programmed confidence trickster is symbolically castrated as a punishment for his Oedipal transgression. William's susceptibility lays him open to deceit and
corruption, quickly drawing him away from the panoramic and toward
the labyrinthine, mythic mode of inhabiting the city (Donald 1995).

William's whiteness in the film thus both promises agency, authority, and total vision and also quickly fails to deliver them, for he both
embodies the law (here, the code) and becomes vulnerable to its transgression through his liaison with Maria. Like so many noir protagonists
before him, William is transformed from the all-knowing, all-seeing
hero to the criminal fugitive on the run, emasculated by love. William is
gradually robbed of the privileges of his racialized white purity when his
desire leads him into the realm of the impure: the narrative structure is
a journey from privileged insider to illegal outlaw. Maria leads him into
the city's demi-monde of forged papers and fraudulent identities, where
William witnesses her giving fake papels to the only black character in
the film, Damian Alekan (David Fahm), who is also the only character
who dies. Pathologized as ignorant, Damian dies as a result of his desire
to travel beyond the inner zone, unaware of his genetic vulnerability
which proves fatal on the outside. These associations with both racialized and legal outsiders incrementally corrupt and pollute the previous privileges of William's purity, coinciding with the impotence of his
empathy virus (his loss of clairvoyant vision), and the expiration of his
papels (his loss of mobility).

Most importantly, this changing sense of William is concomitant
with the revelation of the couple's violation of code 46 and thus of the
impurity of their desire for each other. William's trajectory toward en-
dangerment (through his association with Maria) moves him increasingly away from his privileges, and it parallels the mounting risk of
their Oedipal transgression: incestuous sexual and reproductive impurity. Like a rewriting of the policing of miscegenation under colonial
rule, code 46 is the violent implementation of appropriately differentiated kinship, echoing old colonial fears about racial mixing through
reproduction, and about the so-called bad blood of impure bodies.
The diminishing potency of William's whiteness is located within this
convergence of the biological and the visible. If the colonial desire to
make racial difference visible—to see its biology through the body and
track its genealogies on its surface—has always been threatened by the
potential of hidden racial mixing, then anxieties about undesirable genetic combinations in the more dispersed networks of globalizing cultures might be seen to rework the uncanny hauntings of colonial modernity. Here the anxieties about the biological visibility of racialized
bodies of colonial modernity mutate into fears of the threat of invisible

genetic kinship in the future (see Gonder 2003). The proximity of these two notions of purity (genetic and racial) refers back to their shared eugenic history, while also complicating their potential future relationship through genetic engineering.[15]

Code 46 requires sufficient genetic mixing (exogamy) to achieve the desirable reproductive purity (a healthy population). Clause 1 of the code, displayed in the opening shots of the film, reads: "All prospective parents should be genetically screened before conception. If they have 100%, 50% or 25% genetic identity, they are not permitted to conceive." In naming the prohibition of incestuous sexuality and reproduction a code, the film combines nature and culture, biology and the law. As a code, the law matches the form of the genetics which necessitates its existence, and nature has mutated seamlessly into culture through a logic that justifies the laws of both. As a prohibition, code 46 legally enacts a conflation that has troubled conceptualizations of the incest taboo more generally. As Derrida has argued in his reading of Lévi-Strauss's (1969) theory of kinship, the incest prohibition is both a biological universal and a "prohibition, a system of norms and interdicts; in this sense one could call it cultural," highlighting precisely the contradiction which lies at the heart of the structuralist project, based as it was on the assumption that this prohibition was the exception that proved the rule of the nature-culture distinction (Derrida 1981, 283).[16] The conflation of the Oedipal with the incest taboo in code 46 rehearses the ways in which genetic engineering belongs to the register of both the natural and the unnatural: it is simultaneously biological and artificial. In placing genetic engineering within a transgression of a law against incest, *Code 46* locates the threat of cloning within a long-standing taboo which collapses the biological prohibition against incest with the cultural transgression of Oedipal codifications.

The full force of the affective disturbance of the film operates through this uncomfortable conflation (had William inadvertently impregnated another relation—his aunt, cousin, or even sister—a different register of affective disturbance would have been at stake). This transgression is amplified through the ways in which the forces of genetic engineering and romance confirm each other, precisely where they should diverge: the incestuous union seems propelled (rather than repelled) by their shared kinship traits and physical resemblances. The couple's romantic connection is given a sense of inevitability through Maria's retrospective voice-over in the opening sequence of the film. Genetic engineering and romantic love reinforce each other through a notion of predetermination, fate, and biological destiny, extending this unsettling affective vision of cloning into the uncanny of genetic temporality.

> The city is the way we moderns live and act as much as where . . . what then does it mean to live in this world where all social relations are reduced to calculation, and yet, at the same time, our experience remains that of a phantasmagoria? How can such a bewildering and alien environment provide a home? (Donald 1995, 81)

The spatial problem synthesized by Donald provides a starting point for considering the troubled temporalities of the genetic imaginary: the promise of genetic prediction offers both an appealing and a terrifying future of certainty. The failure of biological transparency to exorcise the uncanny not only presents genetic engineering as an imagined but impossible solution to the mapping of internal and external spaces (whose structural threat to authenticity and singularity requires the exclusion of undesirables), but it also locates the genetic uncanny as a disturbance to temporality. As chance, risk, and unexpected outcome are replaced by predetermination, predictability, and instrumentality, the temporality of the human is redesigned by the desire to collapse linear time and bring the present into the future (to offer a crude summary of the temporal disturbance at stake here). Just as the spatial calculation which promises clarity of vision for modernity functions like the clairvoyance of genetic prediction for postmodernity, so the phantasmagoria of inhabiting the dream world of the modern city gives spatial form to the nightmarish possibilities posed by the scrambled temporalities of genetic engineering and cloning. Such fantasies of both spatial and temporal control require the repression of the unknowable other within, which returns in *Code 46* as the uncanny hauntings of Oedipal kinship. What we might call the warped temporalities of genetically engineered kinship register these temporal disturbances, which do not simply compress linear time (bringing the future into the present through prediction) but are accompanied by the potential shattering of the linear and sequential time frames of biological inheritance, genealogy, and relatedness.

Drawing on Lacanian conceptualizations of the uncanny, we might shift to a formal, structural level. As Strathausen argues, unlike the

Freudian notion of the uncanny, the Lacanian conceptualization of the uncanny "is not bound to an event or a tangible object, but to a crisis in perception, which jeopardizes the integrity of the body and the identity of the subject" (Strathausen 2003, 28). According to this formulation, the uncanny might be thought of as a kind of structural disturbance that emerges with modernity, "which prevents it from ever being fully present or completely visible" (ibid., 16). Mladen Dolar identifies the uncanny as "constantly haunt[ing] [modernity] from the inside," as a symptom of "the *real* shattering the *apparent* wholeness and controllability of social reality" (Dolar 1991, emphasis in the original, quoted in Strathausen 2003, 15–16). The uncanny here is thought of as that which "distorts an otherwise perfect picture," literally functioning "as the blind spot of modernity's preoccupation with visual transparency insofar as it must necessarily remain obscure in order to enable vision" (Strathausen 2003, 16). Following these formulations, "any attempt to disclose the uncanny or to exorcise it by making it visible is bound to reproduce it," since "the Lacanian uncanny is a void signifying the implicit limits of interpretation, the lack of meaning necessary to sustain it" (ibid.).

Code 46 articulates the structural disturbances posed by the uncanny temporalities of genetic engineering through an Oedipal romance narrative which maps the divergent modes of inhabiting the city onto their traditional associations with sexual difference. William's corruption by Maria in the film is marked by his descent into the phantasmagoric spaces of the city, inhabited through a tactile, embodied encounter as opposed to an aerial, controlling gaze. While the uncanny functions as "the blind spot of modernity's preoccupation with visual transparency," its configuration as a forbidden Oedipal narrative in the futuristic control society of *Code 46* distorts the "otherwise perfect picture" of genetic clairvoyance which promises a predetermination that will bring the future into the present. Highlighting the limits of the perfect vision of genetic predictability as "apparent totality of social reality," this uncanny romance disturbs the linear temporalities which genetic transparency might have ensured. In other words, just as the uncanny has represented the "internal limit of modernity, the split within it" (Donald 1995, 82), so here it stands for the threat to the subject in a world of cloned fetuses where the temporal laws of Western kinship have lost their foundational principles: linearity, continuity, and generation (see Franklin and McKinnon 2001). When William has sexual relations with

Maria, he is simultaneously having them with his mother, sister, and lover.[17] These indeterminacies of the scrambled temporalities of genetically engineered kinship (someone could be your lover, mother, and sister simultaneously) disturb the traditional flows of modern temporality, producing a new, uncanny haunting of the desire for total vision (see chapter 9).

It is Maria who most acutely embodies these temporal disturbances, and it is her body which eventually bears the violence of its impossible resolution. The film's use of Maria's retrospective voice-over re-enacts the uncanny as the limit to the desire for omnipotence and ubiquity. With the authority of the invisible narrator and the wisdom of hindsight, this voice-over has a ghostly, predictive presence, like a vocal mimic of the promised clairvoyance of genetic determinism that William embodies through his empathy virus. Maria's voice-over at the beginning of the film (addressed to William like a personal letter after the end of a relationship) accompanies scenes of him in her absence (such as the opening shots of his arrival in Shanghai discussed above). Maria's voice-over promises an all-knowing perception within the film, but her narrational authority is disrupted by the discontinuities in her own knowledge and the disturbances to her memory and subjectivity, resulting from the surgery following her violation of code 46. The sense of narrational control afforded by her voice-over has an increasingly disorienting feel, once Maria's memory of the affair has been removed. Scenes in which Maria is clearly not diegetically present are haunted by this sense of temporal and spatial dislocation: in William's home in Seattle, for example, a fleeting shot of Maria in the corridor gives her ghostly presence in his family life on the other side of the world. In the final scene of the film, her voice-over returns to tell us that her confiscated memories have been restored by the authorities as a punishment. Like a metonymic enactment of the uncanny of genetic prediction, the tension between the omnipresent narration, with its wisdom from hindsight, and the increasingly evident gaps in Maria's own diegetic knowledge produce a sense of temporal discrepancy for the spectator. If the uncanny represents "a crisis in perception which jeopardizes the integrity of the body and the identity of the subject," as Dolar suggests, then Maria's disintegration through body modification and prosthetic memory surgery interferes with the temporal coherence of her narrational account (Dolar 1991, quoted in Strathausen 2003, 28).

The clinic scene on the outside is pivotal to this reading of the film, since both Maria and William lose something crucial to the continuity in their sense of subjectivity, which is so central to their romance. It is also the crucial scene in terms of the suspense of the romance narrative: will they or won't they overcome the cumulative barriers to their love for each other? Their parallel losses transform their relationship to time and to each other: Maria loses her memory of their affair and her pregnancy; William loses both Maria (since she no longer remembers him) and, importantly, the power of his empathy virus (since it is tied to the spaces of the inner zone).[18] With this loss, his ontological security status changes, and he is shown as the object of surveillance (rather than its agent) in the long shots—in the style of closed-circuit television—of him seeking access to the clinic, emphasizing his isolation and impotence. From this moment in the film, the previous sense of William and Maria's shared, temporally and spatially coherent romantic scenario gives way to an emergent mistrust of the potential temporal distortions of technological intervention: the unity emphasized by coincidence and similarity is replaced by a discordant divergence of memory. The loss of their common history and perception, so central to the narrative drive of the romance, undermines the continuation of the spectator's desire for their union (the film arguably never regains this narrative drive and fails as a romance at this level). Everything, it seems, can now be artificially produced, implanted, and removed: history, desire, empathy, and memory. The instant, impermanent, and artificial temporalities of the control society of genetically engineered bodies reorder continuities and authenticities that had been taken for granted: desire is authentic one minute and artificially removed the next; an empathy virus gives mind-reading capacity in one space, but not in another; a prosthetic memory of one kind of surgery wipes out the trace of another.

The temporal disjuncture in Maria's postoperative screen presence and voice-over narration produces a sense of the uncanny of genetic temporalities: the inevitable failure of the desire for temporal ubiquity. Maria's voice-over narrates a love story of which she has no memory, until its ultimate restoration with the narrative closure. Maria and William's mutual "crisis in perception," to use Strathausen's phrase, coincides with a transformation in their embodied relationships to technologies of transparency. Thus, as Maria loses her integrity as the narrational source of the diegesis, so William loses his integrity as the embodiment (and human face) of technological surveillance in the genetically engineered future, and he ultimately loses his memory of the whole affair. The interference with their memories of desire for each other, in accordance with the regulations of code 46, disturbs the boundaries around the subject and the coherence of the temporalities of their shared personal histories.

The temporality of this uncanny Oedipal romance is configured through a series of dreams, hauntings, premonitions, coincidences, and mythic repetitions that produces a disquieting sense of their romance as a biologically scripted, but forbidden, fate. When Maria is first seen in the film (giving the anonymous voice-over a physical presence within the diegesis), she is shown reenacting her recurrent birthday dream: she searches a subway train in slow motion, intently looking for someone she must find before she wakes up. "Each year there is one less station to pass before I wake up . . . It began with 18 stations, and this year it is down to one," her voice-over tells us. The introduction of Maria through her obsessional search for a lost—or not yet found—person in her dream gives the spectator a sense of her interiority. As Charlotte Brunsdon has argued, in her study of the cinematic spaces of the London Underground, "underground spaces can be particularly amenable to the project of suggesting interiority" (Brunsdon 2006, 13). She suggests that the Underground has been "a space to which the lack of an outside or a view lends a curious privacy . . . it is also [a] . . . narrative space which prefers certain kinds of stories about the forgotten, the repressed, and about pursuit" (ibid., 17). Maria's remembered dream maps out the path of her romantic destiny through the subway train, like an unconscious wish fulfillment grafted onto the clairvoyance of genetic prediction (this is indeed the space where Maria initiates a romantic encounter with a stranger). Following Benjamin, Vidler argues that the subway is the equivalent of a city's unconscious: "buried in the subterranean ways of the modern city, was the figure of the labyrinth, site of endless wandering" (Vidler 2000, 75). Like the unconscious of the city, the subway in *Code 46* functions as the mise-en-scène of Maria's dream world, as the site of her "endless wandering" in search of her romantic and genetic destiny. Like the uncanny disturbance to temporality of genetic prediction, this dream narrative prefigures the time and place of her finding her soul mate, William, not yet known to be too closely related. Governed by a sense of destiny, both in terms of her unconscious drive (toward the one who was meant for her, the only one) and in terms of the coincidence of a meeting between such a geographically distant but closely related couple, William and Maria appear to be heading for one another. Symptomatic of the impossible transparency of genetic predictability, this relationship is both against all the odds and yet meant to be, risking everything and yet predetermined.

The film builds a sense of the temporal discrepancies at stake in the genetic uncanny through visual prefigurations of the genetic match, which are only retrospectively given narrative logic. The sense that William and Maria's mutual fascination lies in some kind of recognition of their uncanny similarities is established in their first evening together, with their common tone-deafness and their strikingly "doughy resemblance" (as one reviewer put it; Hoberman 2004). The repeated mirroring of profile and close-up shots of each of them emphasizes their shared facial features: white, fleshy cheeks; wide mouth; squidgy nose; and blue eyes (see figures 26, 1 and 2). The uncanny feel not only of their physical resemblance, but of their narcissistic attraction to each other because of, rather than in spite of, their genetic similarity works to produce a disturbing sense of the erotic charge of their tied genetic fate. Far from working to repel the couple (as required by the incest taboo), their genetic similarity seems to produce sexual desire. The couple's mutual physical attraction is disturbingly placed within the frame of their physical genetic match so that their warped kinship increasingly comes to define their desire.

Maria's tomboyish, youthful style and cunning reserve also echo William's older, professional embodiment of the same characteristics (at the moments when he's cheating the security systems, he also seems boyish and enigmatic). Their shared boyishness is part of a dynamic between them, which is full of homoerotic associations: the younger "boy" picks up the older businessman in a public space and seduces him in an unfamiliar city. Maria's appearance in her black cap and coat, with cropped hair and urban self-confidence, also places her within an aesthetic of metrosexuality, with a distinctly queer urban feel (see figure 27). These homoerotic connotations are extended into the filming of the lovers' first sex scene, which cuts between the two evenly, emphasizing not sight but touch in the close-up shots of their hands and mouths. Such mutuality and visual symmetry is rare within the scopic

26. A romantic attraction based on genetic sameness: the physical resemblances between (1) Maria Gonzalez (Samantha Morton) and (2) William Geld (Tim Robbins) in *Code 46* (Michael Winterbottom, 2003).

economies of penetrative heterosexual sex on the cinema screen, and the scene emphasizes similarity and reciprocity.[19] Despite its vigorously heterosexual narrative drive, *Code 46* thus nevertheless refers to the associations of cloning with narcissistic homosexual relations (see chapters 1, 2, and 5).

Anxiety about the place of sexuality and reproduction in a world beyond Oedipus here connects to the problem of the future of the cinema, if the structure of this foundational myth—which has been a mainstay of films' visual and narrative pleasures—is threatened with fracture (Mulvey 1989). What could replace the generative impetus of Oedipus if kinship were profoundly restructured by genetic engineering and cloning? The incestuous transgression of the couple in *Code 46* shifts their erotic dynamic from the reciprocity of homoerotic equivalence to the violent imposition of heterosexual domination. In this move, the pro-

27. The boyish metrosexuality of Maria (Samantha Morton) in *Code 46* (Michael Winterbottom, 2003).

hibition against homosexuality precedes the incest taboo in establishing the boundaries of proper kinship (Butler 1990 and 2000).[20] The full affective impact of the warped temporalities of genetically engineered kinship condensed in the Oedipal transgression of code 46 is delivered in the film's second sex scene, which can only be described as an act of consensual rape. To overcome the "anti-William virus" implanted in her by the authorities, Maria invites him to force her to have sex with him, so the sense of discomfort at her violation is undercut—and there is also the notion that the sex is a form of resistance against oppressive state regulation.[21] In the world of authoritarian genetic determinism, where secrets are such prized commodities, William's continued silence on the subject of their biological relatedness can be read as both complicit and resistant. The scene's power to generate a haunting and lingering sense of the genetic uncanny operates through the cumulative force of multiple sites of disjuncture. The discrepant knowledge of the two protagonists (he knows—but she does not—that she shares his mother's genetic identity) exacerbates the discomfort of witnessing the violation by underscoring the bifurcated spectator position enacted through their embodiment of competing sexual responses. Contrasting sharply with the mutuality of the couple's earlier sexual encounter, this second sex scene has a disjunctive force based on the violent disruption to temporal continuity produced by Maria's genetically engineered anti-William virus. In his attempt to return Maria's body to its previous state of desire, William restages the mise-en-scène of their earlier sexual encounter, encouraging Maria to sing the same song she sang to him before they had sex in her apartment (as if the singing body might prompt her to remember the desiring body). Since their mutual tone-deafness is one of the signs of their inappropriately close genetic kinship, and thus of the incestuous nature of their desires, its reiteration as a prelude to sexual transgression places not only his physical force (he ties her down to the bed) but also his lack of disclosure (of their relatedness) within the frame of this violent sexual encounter.

The scrambled temporalities of genetically engineered kinship in this scene produce an acute discomfort for the spectator as the couple endeavors to bring something dead back to life, to retrieve their lost mutual desire. Maria's embodiment of these conflicting temporal registers is shown in the discrepancy between her prosthetic repulsion (which even to the most moderate of feminist spectators registers as the reaction of a woman being unwillingly penetrated by a man) and the close-up shots of her face mouthing to William the words "I love you" (she thinks her desire but cannot feel it). The close-up shot of Maria's face filling the screen as she orgasms, as if such physical intensity might eradicate other ghostly hauntings, suggests that authenticity triumphs and that the past has been restored. Ultimately, though, authenticity here is not opposed to artificiality, for the latter is present on both sides of the axis: the genetically engineered foundations of their original sexual attraction are invoked to battle with Maria's genetically programmed repulsion.

Moreover, these uncanny genetically engineered temporalities stretch beyond the discrepancy between Maria's mind and body, or between her will, memory, and programmed response. It surfaces here in the return of another past which haunts this event (and indeed the whole narrative): the lost maternal body of posthuman reproduction. William's violent penetration of Maria's unwilling body, through which her repulsion is transformed into ecstasy and her repressed (or stolen) desiring subjectivity is returned from the past, is simultaneously the literal enactment of his desire for his mother's lost body—the return of the repressed of an Oedipal kind. The "crisis in perception which jeopardizes the integrity of the body and the identity of the subject" proliferates here as the spectator witnesses these simultaneous, multiple transgressions (Strathausen 2003, 28). The lost maternal body retrieved here is both William's own mother and the more general loss of its capacity to signify definitive origins and linear genealogies through genetic engineering and fetal cloning. If the maternal body signifies the anchor of modern temporalities, its symbolic loss here registers the scrambled time zones and warped kinships of genetically engineered futures.

To complete the Oedipal circuit, the other ghost in the second sex scene is Maria's father. As Maria tells William, her father "used to tell stories about Jebel Ali and the Arabian sea; he used to live there and believed if we could return there everything would be all right." In this sense, the couple's flight to Jebel Ali is a nostalgic return to the place of the lost father, a vicarious enactment of the father's homesickness by the daughter.[22] The couple end up in a sprawling, sandy town of low buildings, with a slow ponderous mood and earthy aesthetic symbolizing the possible return to a previous era of traditional, low-tech immobility: here there are no *papels*, no security technologies, no surveillance cameras—only old-fashioned telephones, worn-out cars, and slow-running showers. This is the other side of the calculated transparency of global megacities. Instead of the smooth surfaces of grey metallic and glass at The Sphinx or Westernworlds—with their geometric, gridlike mappings of anonymous spatial zones and their clinical styles of blue, gray, and white interiors—this is a place of hospitality on the outside, full of darker, richer colors and rough, unfinished textures of stone and wood. This is the space of slow time, contemplation, community, connection, and cooperation, where the couple is offered "a room, food, a game of chess?" This is a return home for Maria, who is William's guide in an unfamiliar place: "Every time I ever imagined this, I always imagined coming here with you even though we'd never met." The return to the outside is a return to the land of her father, which represents everything lost in the high-security zones of global urban speed, mobility, and transparency.

If the second sex scene is a retrieval of Maria's eradicated desire for William, a desire which the film established to be based on their genetic familiarity, then the return of repressed desire through repetition also resurrects their lost parents, whose presence is invoked through a sense of the past inscribed in these contrasting interior and exterior architectures. While *Code 46* relies upon that most cinematic of formulas (a son unwittingly has sex with his mother), it is an Oedipal drama about genetic engineering and cloning that undoes the conventional temporal sequence and directional flow of the narrative conventions upon which it also depends. For the (failed) Oedipal attachments here (William to his mother, and Maria to her father) raise disturbing questions about the meaning of biological security in a future of genetically engineered kinship, questions which are hard to resolve in the narrative's own terms. As the cloned fetal twin of William's mother, in what sense is Maria biologically related to the father figure whose desires to return home she enacts? The uncanny possibility of inhabiting the same generational temporality as one's own parent undoes the teleogical flow of the classic Oedipal trajectory, and it compresses generational sequence through genetic engineering to produce new forms of kinship dislocated from the system of blood ties. The traditional resolutions of sacrificing the other woman in the interests of heterosexual romance, or banishing the femme fatale who has brought about the downfall of the innocent male protagonist, fail to resolve this particular version of the Oedipal conflict.

A film about the scrambled temporalities of the warped kinships of genetic engineering and cloning, *Code 46* undoes the previous coherence and continuity of modern embodiment, however illusory or fragile. A post-Oedipal scenario located within the postcolonial landscapes of past and future technologies, the film ties the potential impurity of white desire to the problem of misrecognition in the changing relationships between visibility and biology in a world of globalized biosecurity. In exploring the question of what constitutes biological life, when the temporal sequences and continuities of kinship no longer hold sway through the competing versions of life on the inside and on the outside, the film places the dubious progress narratives of Western technoscience and the multicultural project of cosmopolitan megacities within the same frame. The artificial intervention into the processes of human reproduction is embedded in the film in the particular practices through which modes of life are designated desirable or undesirable, human or not fully human, according to their place within the narrative of technological development.

In *Code 46*, lost ways of life on the outside ensure the effectiveness of regulation in the inner zones of the future. If Wollen is right that we live in a culture that has come to value "mobility over memory," then Maria's punishment and William's return home may be symptomatic (Wollen 2002, 214). In the final scene of the film, we see Maria in exile on the outside. As Vidler (1992) argues, the undesirable and much feared zones of exile are as fundamental to urban authority as the desired transparency and order of the modern metropolis. The structural dependency of inside and outside spaces organizes this final scene, which cuts between Maria's wandering in exile and William's recovery in a hospital in Seattle, where he is reunited with his wife and son. Having had his memory of the affair removed, following the clichéd gendered injustices of romantic narrative, William is able to have it all (a passionate affair, followed by renewed desire within his marriage) while Maria is left with nothing but her haunting memories of a lost love (in this sense, she has been punished for her transgression like so many femmes fatales before her). Here the punished body of the woman represents the failure of whiteness to retain both its purity and its privilege. Ruined by the sexual impurity of an Oedipal transgression, she is cast out of the urban space of healthy (genetic and ethnic) mixing and banished to regressive nomadism, where she inhabits the space of the other: her black, urban Western clothing has been replaced by a loose pale patterned sari, which she uses to cover her head in the sandy breeze. Now she is both ethnicized other and alone in her impure whiteness: the single figure of exceptional white suffering, abandoned to a life meant for others (see figure 28).[23] Having lost the privileges of whiteness, Maria is expelled to the margins of this third-world landscape of underdevelopment and poverty. Having transgressed the genetic laws required for fetal cloning, she inhabits a space devoid of all signs of urban comfort. Going against the laws of nature and culture, she has nowhere to hide, and she walks alone, exposed to the elements.

While the film explores the dangers of the end of Oedipus in a world of fetal cloning, it simultaneously enacts an Oedipal return in the couple's flight to the outside, which makes Maria's exile seem like an inevitable repetition. Just as Benjamin suggests that the city is a series of semblances "where we live once again oneirically the life of our parents and grandparents" (Benjamin quoted in Vidler 2000, 77), so Maria treads again her father's path of exile during her childhood. As Maria now inhabits the endless ground of the outside upon which William gazed down from his airplane in the opening shots of the film, the sense of repetition, so central to the haunting presence of the uncanny,

28. The expulsion of the femme fatale: Maria (Samantha
Morton) as an isolated nomad outside the city at the end
of *Code 46* (Michael Winterbottom, 2003).

collapses the psychic and the genetic through their common mappings
of inheritance and genealogy. Maria is both at home and not at home in
the dreaded space of the outside to which she has been so nostalgically
(and abjectly) drawn through her attachment to her father's narrated
past, despite her better judgment. If it is Maria's genetic and romantic
destiny to fall in love with William, her expulsion to the outside reiterates her father's exile with the weight of genealogical inevitability.

Maria's preoccupation with the outside, expressed early in the film,
appears to have been a premonition at the moment of narrative closure.
The closing shots of her wandering return the spectator to the mood of
her oneiric presence in her narrated dream of the subway at the beginning of the film. Now the dream has become the nightmare, and her
empty mode of inhabiting the space looks suitably haunted. Now she
has lost the one who embodied her destiny, and she wanders the outside,
with what Benjamin calls the "blank stare of the somnambulist" (Benjamin quoted in ibid., 76). The return of her voice-over narration reminds
us that this is supposed to have been her story all along, but this bid for
authorial control fades against the almost hallucinatory feel of her narrational agency. At this moment, Maria finally possesses full temporal
vision since her memory has been restored, yet the trauma of the events
has produced a kind of "ambulatory automatism" (ibid., 74). For Benjamin, neurasthenia leads to vagabondage, and in this final scene, Maria

29. A mise-en-scène of reflective surfaces shows
William (Tim Robbins) alone at the airport in *Code
46* (Michael Winterbottom, 2003).

wanders the spaces of exile like a vagabond, a person who is not fully
human, with "no home, no work and no master" (Benjamin quoted in
ibid., 75). As Vidler writes, "the vagabond . . . alone, criminal and ex-
iled . . . possessed the marginal vision that transgressed boundaries and
turned them into thresholds, a way of looking that engendered what
Benjamin called 'the peddling of space'" (ibid., 73). The final scene of
Code 46 leaves us with a sense of the terrors of wandering. The embodiment of the uncanny at a structural and symbolic level organizes
the spatial patterns and temporal sequences of *Code 46*: Maria now
wanders barefoot in the kinds of sandy spaces William flies over. In his
exploration of the terrors of wandering, Benjamin suggests that "only
a dreamlike state of suspension might enable the wanderer to cross
between physical surroundings and their mental contents . . . viewed
through these lenses, the urban street regained something of the original terror of the nomadic route" (quoted in ibid., 75).

  Code 46 cautions against the desire for total spatial and temporal vision. The fantasies of control through a God's-eye vision of space at the
beginning of the film extend into its exploration of temporal modalities
which allow the future to be predicted: synchronic ubiquity becomes
diachronic omnipotence. Just as the desire for spatial transparency
eluded the architectures of urban modernity, so the predictive temporal totalities, which promise to bring the future into the present based
on past patterns, disavow both the impossibility of such ubiquity and
the terror of an ontology governed by such certainties for the subject

of contemporary genetic cultures. When Maria's voice-over asks the hypothetical question in the film—"If you knew the outcome to every action, how could anyone ever do anything?"—she poses the problem of genetic determinism and prediction thus: is genetic predetermination unbearable to a human psyche so wedded to the notion of agency (however compromised, however illusionary)? The film leaves a sense that the desire for such transparency belongs to the delusionary megalomania of a global state whose authority lies in the violent implementation of a universal genetic law. As we have seen, such fantasies of visual control of spatiality and temporality require the repression of the unknowable other within, which returns here in the form of an uncanny haunting. Like the shiny surfaces throughout the mise-en-scène (see figure 29), this world which promises transparency may end up simply reflecting back to us an illusory image of ourselves which hides a deeper secret.

[3]

Stairway to Heaven

[7] Cut-and-Paste Bodies

The Shock of Genetic Simulation

Is geneticization to the body what digitization is to the image? Both are metonymic relations with the capacity for endless reproducibility through replication. If this analogy is worth pursuing, how might we think about the possibilities of techno-scientific interference in bio-genetic processes as inaugurating a sense of what we might call a lost bio-aura? Since our cells are now thoroughly codifiable as genetic information—which can be tagged, extracted, transferred, reprogrammed, and recombined—and our reproductive capacities can now be amplified, assisted, manipulated, substituted, externalized, or blended with laboratory techniques, previous notions of the sacredness of life, the distinctiveness of the human, and the singularity of embodied subjectivity can no longer form the foundations of modern subjecthood as they once did.

For W. J. T. Mitchell, the processes of genetic engineering are not only analogous to but also inextricable from the transformation of the status of the image in the digital age. He argues:

> Biocybernetic reproduction has replaced Walter Benjamin's mechanical reproduction as the fundamental technical determinant of our age. If mechanical reproducibility (photography, cinema, and associated industrial processes like the assembly line) dominated the era of modernism, biocybernetic reproduction (high-speed computing, video, digital imaging, virtual reality, the Internet, and the industrialization of genetic engineering) dominates the age we have called postmodernism. (Mitchell 2005, 318)

For Mitchell, the changing significance of images as "vital signs" demands that we now consider them as animate objects with desires, needs, and appetites: in short, "images are like living organisms" (ibid., 11). He suggests that the "curious twist in our time" is that the "digital is declared to be triumphant at the very same moment that a frenzy of the image and spectacle is announced" (ibid., 315). In a similar vein, Raymond Bellour's theory of the changing status of image cultures uses the metaphor of the double helix to conceptualize the twists and turns of "on the one hand, more and more differentiation, and, on the other, a virtual indifferentiation" (Bellour 1996, 177). Both Bellour and Mitchell read the new image species of our time through biological analogy, and indeed through the animating power of genomics.

As I have argued throughout this book, in thinking about the cinematic life of the gene, two related cultural phenomena currently undergoing widely debated transformations share the same frame: the cinema and the body. Using Benjamin's famous 1936 essay "The Work of Art in the Age of Mechanical Reproduction" (1969), known as his "Artwork" essay—as a prompt to think about the impact of scientific interference in genetic processes on contemporary modes of perception of the body, I consider the geneticization of the body in relation to modern techniques of photography and cinema and to more recent digital techniques.[1] At its most general level, the geneticized body pushes in both these directions simultaneously: its new modalities present a shock, arguably comparable to the ways in which, according to Benjamin, photographic and cinematic technologies of reproduction led to mediated human relationships to culture; and the geneticized body's fragmenting and disembodying effects on the connections between sexuality and reproduction parallel the digital disturbance to the authenticity and integrity of the mechanically produced image. The concept of the geneticized body is used here to indicate that, since the decade of the clone,

marked by the completion of the Human Genome Project and the cloning of Dolly, how we think about our reproductive bodies has been profoundly transformed by the significance of potential techno-scientific
interference at the cellular level. The more general shift from a mechanical to an informational vision of the human body (see Stacey 1997)
is now anchored firmly within the notion that the gene has become the
fundamental unit of life in popular and scientific discourse. Since we
are now aware that this unit is open to techno-scientific manipulation,
this constitutes a profound disturbance to our previous modes of corporeal perception. In this final short theoretical speculation, I explore
the potential convergences between the ruptures in our perceptions of
image and body introduced by new modes of cultural and biological
reproduction, those of digitization and geneticization.

What might it mean to claim that the geneticized body could generate a sense of lost bio-aura? As an interference with the body's reproductive processes, genetic manipulation might be thought of as
the contemporary biological equivalent to the shock of modernity described by Benjamin in his "Artwork" essay and elsewhere. By introducing technical interference with reproduction into the debates concerning Benjamin's claims about the loss of aura of an art object, we might
think about the convergence of the biological and the cultural in this
detraditionalization of reproduction: how might theories of the history
of cultural reproduction, such as the emergence of photography and
the cinema, shed light on the disturbances to cultural perception currently posed by genetic engineering? How might we think about the
reconfiguration of the geneticized body through the notion of a loss of
bio-aura that has begun to alter our psychic horizons? I define bio-aura
here not as an individual possession, as something one does or does not
own, but rather through the ways in which its loss indicates a changed
mode of perception of human reproductive and sexual embodiment.
Through a dialogue with critical debates about Benjamin's work on the
loss of aura in the age of mechanical reproduction, this chapter discusses the shift in perception occasioned by the new possibilities of
genetic engineering and cloning. Benjamin's highly interdisciplinary
essay asks us to consider how the transformation of the human-technical
interface might produce new modes of perception; it provokes us to
take notice of cultural conventions before they slide into the invisible
of the everyday. My suggestion here is that at this moment, the genetic
imaginary demands a similar perceptual interrogation. Following the

psychoanalytic inflections of chapters 1 and 4, this chapter moves the notion of the genetic imaginary into the frame of disturbances to cultural perception, asking questions about affect, shock, and attachment from a rather different—though related—perspective.

The troubling impact of genetic engineering and cloning on our psyches might be conceived as a shift from the authority of the original body to the technological reproduction of numerous copies. Conceptualized thus, this current move parallels Benjamin's claim that modern technologies of reproduction transformed our relationship to art, from one based on a sense of its presence in the here and now to one based on a perception of its artificial replication—resulting in what Benjamin famously calls the loss of aura. In exploring this move from authentic singularity to artificial duplication, I take his analysis of changing modes of human perception and art forms into the realm of genetic engineering and cloning. Extending Benjamin's concept of the loss of aura to the domain of the geneticized body, we might think of the demise of bio-aura through the fading sense of the body's singularity, nonrepeatability, uniqueness, integrity, and authenticity. As Mitchell puts it:

> *Reproduction* and *reproducibility* mean something quite different now when the central issues of technology are no longer "mass production" of commodities or "mass reproduction" of idealized images, but the reproductive processes of biological sciences and the production of infinitely malleable, digitally animated images. (Mitchell 2005, 318, emphasis in the original)

To conceptualize the shock of genetic interventions through the use of the notion of the loss of bio-aura suggests something of bio-aura's affective force, its perceptual disturbance, and the depth of its reach. For Benjamin, to be human is to sense affect. The threat of its loss distorts our connections to others and the objects around us. Bio-aura, it would follow, concerns the embodied relationality of our future, and its loss would be signaled by the disappearance of that genealogical connection. In a world of genetic engineering and cloning, human presence no longer connects aura to affect. What Teresa Brennan (2004) calls the "transmission of affect"—the capacity to sense others and be moved by their intersubjective presence—forms the ground for the dynamic production of embodied subjectivities. Here, as elsewhere in affect theory, the biological and the emotional are inextricable. But what happens to the transmission of affect in the world of genetic engineering, especially when Brennan's theory rests on an in utero formation of affective perception? If bio-aura relies upon the transmission of affect between embodied subjects, what new relationalities arise when those bodies (or components thereof) have been artificially engineered? How might we feel a sense of the loss of bio-aura in ourselves and in others, or in our relational exchanges? Is this the kind of shift that enables a theorist like Eugene Thacker (2004, 31) to slide from the human to the micro-biological and ask "what is biomolecular affect?" (see chapter 8).

If the word "aura" can be understood as an affective and present relational connection between bodies and artifacts, bio-aura might be thought of as a sense of the transmission of humanness based on gene-alogical, integrated, and unmediated vitality. As successful imitations of human reproductive life, genetic engineering and cloning threaten the previous sense of humanness located within a particular intergen-erational capacity for generation, simultaneously able to initiate new life and to avoid or postpone death. The threats to bio-aura posed by genetic engineering and cloning concern the potential for technical manipulation of the cycles of life and death—scrambling generations and toying with immortality. Similarly, for Benjamin, portrait photo-graphy—in which the subject continuously looks into the camera but finds no return of his or her gaze—is inhuman, even deathlike, running, in Richard Shiff's words (2003, 67), "counter to the experience of aura." By extension, as the inhuman counter to bio-aura, genetic engineer-ing threatens to taint human reproduction with a loss of authenticity, transforming our perception of the life-giving processes of the human body into a set of scientific techniques in which the promise of life is haunted by a deathly presence.

Benjamin's notion of aura, a word he defined in a number of different ways in his writings, is perhaps one of the most widely debated con-cepts in cultural theory. One of its many definitions, particularly well-suited to a discussion of the changing status of the geneticized body, is the medical one that refers to aura as the precursor to a major epileptic attack: "a curious sensation of a cool or a warm breeze (*aura epileptica*), which, starting from one end of the body, passes through the same, and ends in the head or the hollow of the heart" (Brockhaus quoted in Weimar 2003, 188). Using that definition, Klaus Weimar suggests that aura (literally meaning a breath of wind) only really makes strict sense in relation to Benjamin's famous passage that speaks of breathing "the aura of a mountain panorama or the branch of a tree on a summer afternoon" (Benjamin quoted in ibid., 189). Elsewhere in Benjamin's "Artwork" essay, Weimar suggests, the word "aureole" (a halo around an entire body) might have worked better than "aura" to designate the uniqueness of something (though he admits that to "inhale" an aureole would be a "bold metaphor"). Citing Benjamin's multiple and confusing uses of "aura" in his essay, Weimar argues that the word does not fit much of what Benjamin seeks to capture: "one cannot really speak of 'atrophy,' of the 'decay' and 'destruction'" of a breath of a wind, nor of its "'being around' someone" (such as Macbeth, or an actor playing Macbeth), and certainly not of its "'shrivelling up'" (ibid.). For Weimar, it is as though a medical doctor had scribbled the word "aura" in the margins of Benjamin's text, and it had somehow made its way into the final version as a kind of Freudian slip (ibid.).

In this medical context, an aura is a physical sensation that passes through the body and establishes a particular sense of presence (here as a precursor to an uncomfortable attack). Beyond medicine, Benjamin's aura has been characterized *inter alia* as: "non-repeatability" (Baecker 2003, 13); "the traditional power of the original" (Gilloch 1996, 53); "an experience of unmediated bodily contact" (ibid., 65); a mode of "contemplation rather than distraction" (Ritter 2003, 209); "something you can sense but not see" and a kind of observation in which the "distance between you and the object dissipates yet it remains distinctly separate from you" (Shiff 2003, 64 and 65). According to Dirk Baecker, aura is not equivalent to uniqueness; rather, uniqueness is "the condition of a possibility of the work's here and now[ness]" (Baecker 2003, 12). The location of aura also shifts in discussions of Benjamin's essay: it is to be found in religion, a work of art, nature, premodern culture, the viewer, and the actor or performer. The loss of aura has been variously designated as regrettable, inevitable, desirable, and potentially revolutionary, and it has been attributed to technological reproducibility, changing relations between viewer and object, and the absence of presence. The ambiguities of Benjamin's use of the word have generated numerous unanswered questions: Is the presence of the art object in the here and now necessarily tied to its singularity and nonrepeatability? Was the authenticity of the real object replaced by a different sense of the authentic in the new cultures of modernity—and if so, was aura already lost, or was it displaced and reinvented? Is the loss of aura more the "memory of religion in art" (ibid., 13)?

Despite all the confusions and disagreements around the notion of aura, Benjamin's "Artwork" essay has been one of the most influential pieces of writing about modern culture. In the reading that emphasizes the loss of aura, this essay makes a move that Raymond Williams (1975) argues is typical of writing about the history of technology, as each new generation locates the loss of something authentic, something unspoilt, something that was more natural in a previous generation. Williams cautions against the projection of a nostalgic longing for nature onto a lost golden age which is now to be mourned. But it may also be that the sense of original presence (of unspoiled, unmediated contact with nature) is itself an illusion, based on an investment in the sacred meaning of objects and artifacts—though this is not to suggest that it feels any less real to the subject once it has been given symbolic form in the cultural landscape.

In their collection of essays on Benjamin's contribution to thinking about the changing status of art in the digital age, Ulrich Gumbrecht and Michael Marrinan (2003b) suggest that perhaps the continuing relevance of the "Artwork" essay lies precisely in its ambiguities and inconsistencies. For Antoine Hennion and Bruno Latour, this may even be the "main source of fascination which the essay . . . continues to exert," stemming, for these critics, from its profound misunderstandings of both historical and modern phenomena (Hennion and Latour 2003, 91). Benjamin's admirably ambitious collage (which Hennion and Latour find to be full of confusions and misplaced predictions) hangs upon the ambiguities of the status of the word "aura," which "organizes not only his argument but also most contemporary discourses on modernity and the past" (ibid., 92). Particularly problematic, they suggest, is the way in which Benjamin's retrospective account "looks back at the modern epoch, [in which] aura becomes a kind of Lost Paradise, a negative foil to what he describes as the new effects of mechanically reproducing works of art, and to the new seductions of the masses that have replaced the former beauty of art" (ibid.). But, they go on to suggest, this is confusingly inconsistent, for when Benjamin turns to the past, he sees the "nostalgia for the aura as an illusion, a relic, or a residue of cult value" (ibid.). Thus Benjamin offers us an ambiguous notion

of aura which Hennion and Latour synthesize thus: "modern standard-
ized copies of art have lost the authenticity of the real presence, but . . .
this now-absent real presence was itself an old religious artefact" (ibid.).
Even if we were to accept that the "modern fetishism replaces God by
idols," Benjamin's "confusions between idolatry, fetishism, art and religion" in this essay, they suggest, leads to "unsustainable claims about
the impact of the cinema on the perceptual formations of the modern
masses" (ibid., 93).

The notion of aura is nothing if not ambiguous, as Hennion and
Latour's compelling critique demonstrates. But rather than taking the
ambiguity of aura as a sign of its failure, I shall use it to explore the
problem of formulating precisely what it is we fear we might lose with
techno-scientific interference in reproduction at the genetic level. For
our shudder at the prospect of cloning (and related techniques) is both
a physical and psychic one, and the threat it poses is both disturbing
and compelling. The landscapes of the genetic imaginary are not neat
or fixed; their shape and formation shift through anxious associations
that register disturbance at the level of perception. Our response to
the new forms of artificial life inaugurated by genetic engineering combines sensation, affect, and unconscious investment. The new relationalities of geneticized embodiment generate a sense of loss of bio-aura
that is hard to name.

Why is it so hard to find language to articulate the disturbance posed
by genetic engineering and cloning? Is it because the difference between
organic and inorganic genetic activity cannot be made visible in a conventional image? Or is it because there is no particular moment we can
identify with the emergence of the threat, nor any distinctive scientific
technique that can be labeled the definitive break with nature, any more
than previous techniques could? These ambiguities make the loss of
bio-aura hard to pinpoint in any self-evident sense. Since biogenetic
techniques dissolve the nature-culture distinction upon which aura relies, it seems that biology has simply taken over from nature without
our being able to see the difference (see Franklin 2007 and Haraway
1991 and 1997). Thacker suggests that the blending of genetic and computer science produces biomediations in which the emphasis is less on
"'technology as a tool,' and more on the technical reconditioning of the
'biological'"; we are moving toward a model, he argues, that is "technically enhanced but still fully 'biological.'" He highlights how "biomedia

consistently recombine the medium of biomolecular systems with the materiality of digital technology" (Thacker 2004, 5, 7). Thought about in this way, the problem of using the notion of the loss of bio-aura to capture the physical and psychic shudder at—or the shock of the perceptual disturbance of—genetic interference might be that to many, it signifies less a lost object and more a process of recharging nature through the biological.

To begin to address this problem, I draw two related hypotheses from the critique of Benjamin's use of the word "aura": first, that the sense of aura is largely defined through its loss; and second, that this sense of loss retrospectively produces something of an illusory or mythical past. If the use of aura in the "Artwork" essay is both illusive and mythologizing, this is precisely why it could be adapted for thinking about the reconfiguration of the body through the new reproductive possibilities of genetic engineering and cloning, which generate a sense of our own peculiarly human bio-aura. Perhaps bio-aura is actually the by-product of its own demise: with its liquidation on the horizon, the body's aura registers an affective presence; with the threat of a technological takeover, bio-aura is produced as a perceptual force.

A related point is that with the demise of bio-aura, a nostalgic longing for an organically reproductive body is projected onto a mythical past that retrospectively constitutes all previous forms of reproduction as untainted by technology, and as belonging to the natural order of an unmediated register (a fantasy of purity that negates the history of medical intervention in pregnancy, childbirth, and gynecology generally). The reproductive body is imagined to have been previously connected to nature through its procreative capacity, authentic integrity, generational sequence, and genealogical lineage. At the moment of disruption, pure biological reproduction comes to symbolize the traditional embodiment of modern spatiotemporal relations. Technoscientific interference in genetic processes represents the end of our embodied sense of integration, distinctiveness, and individuality, ideologically changed as that perception may have been.

However imprecise (perhaps even through such imprecision), the mourning of a lost bio-aura seems to have some kind of centripetal effect within the genetic imaginary in unifying multiple forms of detraditionalized reproduction. As we have seen in the theoretical writings and films discussed so far in this book, the genetic imaginary draws high-tech scientific interventions, such as cloning, together with a wide variety of infertility treatments, including in vitro fertilization and egg and sperm donation, as well as low-tech alternatives to heterosexual procreation, such as self-insemination and surrogacy. Although only some of these involve scientific techniques in laboratories, their challenge to cultural norms is cumulative and expansive. In the context of multifaceted disturbances to what Judith Butler (1990) famously called "the heterosexual matrix," a sense of the loss of bio-aura is produced through a complex relay of affective forces. In uncoupling sexuality and reproduction, cloning takes a central place on an already crowded stage of detraditionalized biologies. What we might call the bio-aura of heterosexual reproduction retrospectively materializes as its own anchor in nature. In the genetic imaginary, this detraditionalization of reproductive heterosexuality is constituted through the affective force of the anxieties governing a lost bio-aura. The sense of a lost bio-aura enacts a form of heteronormative nostalgia in the phantasmagoria of new modes of reproductive and sexual replication.

Debates about the digitization of image cultures have returned many scholars to many of Benjamin's concerns discussed above. We might sum this up as follows: just as the "here and now"-ness of human relationships to artifacts was transformed by the new possibilities of mechanical reproduction (photography and then cinema), so the indexicality of the image (representing something which was there but is not any more) has in turn been displaced by the new imitative possibilities of digital manipulation, whose claims to a "having been there"-ness are no longer to be trusted. In this context, the genetic imaginary reconfigures notions both of liveness and aliveness, pushing the limits of our investment in the authenticity of images and bodies (see Mitchell 2005).

As the shift from analog to digital imaging techniques signals the potential demise of the distinctive artistic forms so central to Benjamin's thesis, critics have been reassessing his claims about the significance of changing modes of human sensory perception for understanding specific modes of cultural reproduction (Gumbrecht and Marrinan 2003). Digital innovation has not only had an impact upon the reproducibility and circulation of the visual image but has also made it newly transformable (subjects can be relocated, faces blended, and temporalities scrambled) through recompositing and morphing, placing the status of the indexicality of the photographic and the cinematic image center stage once again. This potential loss of immutability has returned scholars to Benjamin's theory of the distinctiveness of the impact of these visualizing technologies, which is so central to his claims about the perceptual frames of modernity and its subjects.

For Lev Manovich, "digital media redefines the very identity of cinema" (Manovich 1999, 173). In his account of what makes cinema a distinctive art form, he suggests that before digitization transformed it, filmmaking—like photography—was defined by its indexicality: "Cinema is the art of the index; it is an attempt to make art out of a footprint" (Manovich 1999, 174). At the moment of its transformation by this new register of artifice (digital mutability), our sense that the cinema has always been defined by its commitment to show us something that has actually been present in front of the camera crystallized. According to Manovich, modern cinema basically shows us things that have occurred in front of the camera: despite a host of techniques shaping the look or style (lighting, different film stock, and particular lenses), and despite the crucial role of editing in producing what finally appears on the screen for the audience, analog cinema nevertheless depends upon "the basic record obtained by a film apparatus" (ibid., 174). Modern cinema is indebted to photographic techniques, and our perception of it has continued to be defined by a sense of lost presence. If "every photograph is a certificate of presence," as Barthes suggests, offering us a simple, but ungraspable feeling of something "that has been," then its difference from, say, writing lies in its ability to "authenticate itself," its power to "ratify what it represents" (Barthes 1981, 85, 87, and 115).

But, as Manovich asks, "what happens to cinema's indexical identity if it is now possible to generate photorealistic scenes entirely in a computer by using 3D computer animation" (Manovich 1999, 175)? What happens to our sense of lost presence—to our investment in the cinema's showing us things that were once present in the time and space of the act of filming—if it is now possible "to modify individual frames or whole scenes with the help of a digital paint program; to cut, bend, stretch and stitch digitized film images into something that has perfect photographic credibility, although it was never actually filmed" (ibid.)? According to Manovich, the "distinct logic of a digital moving image" brings with it the loss of any sense of authenticity, realigning the cinema's relationship to presence: "this logic subordinates the photographic and the cinematic to the painterly and the graphic, destroying cinema's identity as a media art" (ibid.).[2]

The loss of an already lost presence is hard to imagine, but it lies at the heart of Manovich's claims about digital media. In this respect, his characterization of our sense of the cinematic echoes Barthes's (1981) account of what is particular to the power of the photograph: analog cinema shares with photography a claim to the authenticity of a once present—but now absent—indexical authority. For Benjamin, the cinema represented the shock of modernity through the loss of an auratic relationship to art objects; for Manovich, the digitization of the cinema undermines the indexical connection between what appears on the screen and what was present in front of the camera at the time of filming. The invention of an illusory presence through digital techniques raises the question of how such possibilities might disturb modernity's modes of perception. The "here and now"-ness of temporal and spatial cohabitation with the art object is replaced by an indexical relation to the "what was there and then"-ness of analog film techniques; with digitization, this sense of seeing what was once present has been replaced by a technology that constructs cinema from a "may never have been there"-ness. If we follow Manovich's argument, it is as though another kind of aura is attributed to analog cinema at the moment of its loss through digitization. The liquidation of indexicality which digital techniques represent for Manovich takes on something of an auratic quality: presence in front of the camera seems to function as an auratic equivalent to Benjamin's "here and now"-ness of an artwork's aura. We are reminded of Williams's (1975) claim that each new generation finds authenticity in what has just been lost through technological innovation.

The extent to which indexicality is the defining feature of analog cinema and how much its loss characterizes the impact of digitization are both highly debatable. In some ways, claims about the loss of indexicality have relied upon a technologically determinist view of digitization, which ignores the aesthetic implications of such techniques (Spielmann 1999). But in other ways, "the disappearance of the indexical dimension of the photochemically-obtained image and the correlative spread of iconic and symbolic imagery, which is easy to manipulate," make little difference to the audience's experience of the film on the screen, since viewers often remain unaware of the technological changes (Willemen

2002, 15). In other words, although production techniques may have been transformed by new digital options, the impact of that transformation on our perceptual frame within narrative cinema proves much harder to gauge. Paul Willemen's rejection of the arguments made by new-media theorists, such as Manovich, might seem overly dismissive, but his caution is nevertheless an important one. The claim of the cinema's loss of indexicality relies heavily on the notion of our transformed perception as spectators: the loss of a sense that what is on the screen was indeed once in front of the camera. Although such a presence may indeed be a digital effect of computer software, it nevertheless typically aspires to the same imitative function. As Siegfried Schmidt suggests in relation to virtual worlds: they "strive, on the one hand, for the most complete reproduction, on the other hand, for a complete liquidation of the real" (Schmidt 2003, 80). The cinema has thus arguably always operated in the realm of the virtual.

In some important ways, Manovich's claims about digital cinema's necessary loss of indexicality seem to ignore the ways in which analog cinema also produces illusions of "here and now"-ness: for example, studio sets imitate authentic locations, and the use of voice-over gives the illusion of a character's presence in the past. In this sense, perhaps there are formal changes to the demands upon the audience to suspend disbelief, but in terms of the contract at stake in the cinematic pleasures of illusion, much less has changed. Moreover, as Willemen argues, it is the editing after the filming that produces a sense of spatiotemporal coherence; thus, indexicality is not predicated upon presence in front of the camera. Additionally, some genres, such as science fiction and action films, may never have aspired to produce the effect of authentic presence in the time and space of the camera (though of course our suspension of disbelief more generally refers to something else). Different historical periods have produced conventions of authenticity which have not lasted, perhaps illustrating Willemen's critique: to take an obvious example, studio backdrops for driving scenes in classic Hollywood cinema of the 1930s and 1940s seem artificial to us now, but they were credible at the time.

These opposing evaluations of the impact of digitization on the cinema result partly from definitional and conceptual disagreements. If we return to the potential analogies between the impacts of digitization and geneticization, we might reframe Willemen's challenge and ask where exactly do we find the disturbance to reproduction that is posed by genetic-engineering techniques. Relatedly, how should we conceptualize the reproductive body before the possibility of these techno-scientific interventions into genetic processes? The notion of a pre-genetic body makes no sense at all since, for biologists, the body has always been genetic. We can ask, though, what kinds of concepts we might use to describe perceptions of the body before it was subject to current techno-scientific interventions. But as soon as we begin to specify the impact of genetic engineering on our understanding of reproduction, we encounter the problem of the consequent renaturalizing of previous notions of embodiment. How might we think about this transformation without producing a mythical past of the naturally reproductive body?

One of the difficulties in answering this question is that these disturbances circulate in indirect and intangible ways.[3] While the perceptual disjuncture between visibility and legibility may be more debatable in the case of digital imaging techniques, the outcomes of genetic engineering's interference with biological reproduction cannot easily be observed. Dolly the sheep—whether live, on film, or in photographs—looks like any other sheep (Franklin 2007; Mitchell 2005). Her early death was caused by her degenerative biological status, but this difference was invisible when she was alive. And the increasingly uncommon multiple births through in vitro fertilization are the only visible indicator of these children's artificial origin. Unlike digitization, geneticization is an ambiguous and elusive change in how we perceive the human.

If the impact of genetic intervention is not visible on the surface of the body, how might we articulate the psychic trouble it undoubtedly poses—and is there any basis for claiming that the digital and the genetic share a perceptual disturbance of our modern spatiotemporal frame, contra Willemen? If digitization offers the potential to reorganize the spatiotemporal coherence of modern image-making technologies, how does the geneticization of reproduction introduce the possibility of reordering the spatiotemporal relations of embodiment and relatedness? Just as "manipulation in the digital means the possibility to simulate, transform, combine and alter any form of the image through computational processes," as Yvonne Spielmann (1999, 135) puts it, so in genetic engineering the cell can be simulated, transformed, combined, and altered. Just as the digital unit of information can be disembedded from one context and reembedded in another (or decontextualized and then recontextualized), so scientific techniques can disembed and reembed genetic information between cells, bodies, or even species (see Franklin, Lury, and Stacey 2000).

Since both genetic engineering and digital manipulation operate through codes, both are open to temporal transformation or acceleration in ways that neither the body nor the film print could have been in their analog or bio-auratic forms. Taking bio-aura to signify a retrospective construction of the body as generative without artificial or technical intervention, producing a sense of authenticity at the moment that it is perceived to have been lost, we might thus seek to establish the common properties of the analog image and the auratic body. Like digital techniques, genetic engineering can now simulate, as well as manipulate, a new sense of what is, and a new sense of the presence of what is not or has never been (Spielmann 1999, 135). As Mitchell has suggested, "a digital image that is a thousand generations away from the original is indistinguishable in quality from any one of its progenitors" (Mitchell 1992, 5). There is a similar potential for scrambling genealogies and disturbing temporal sequences in a genetically engineered or cloned fetus so that it could be both sister and daughter of its mother. Temporality is accelerated, compressed, and distorted through genetic interference, producing the possibility of embodying two generations simultaneously, blending the traditional genealogical teleologies of Western kinship. In Dolly's case, time is both frozen and accelerated: she is of her mother's generation of sheep, yet she died before her mother because of this (see Franklin 2007).

These temporal disturbances are transformed into an altered set of spatial relations in both cases. As Spielmann suggests, "spatialization is the preferably expressed feature in digital aesthetics" since "what constitutes linear continuity in the first place is transformed from a temporal to a spatial category of connecting elements" (Spielmann 1999, 140 and 137). In discussing collage (see chapter 9), she argues that with regard to digital cinema's spatialization, this means that temporal features are not considered to be simply dissolved but rather transformed, reworked, reshaped, and finally changed in directionality (ibid., 140). Writing of digital aesthetics, Spielmann argues that the digital might be thought of as nondirectional. Insofar as the analog image represents "directional characteristics," digitally encoded images generally have "omni-directional features" (ibid.). In genetically engineered reproduction, the body is also respatialized through techniques such as nuclear transfer, cell implantation, or in vitro fertilization. When the biological body is turned into a reproducible copy, a sense of the body's aura is lost forever. Undoing traditional teleological lines of descent, these techniques produce potentially chaotic, unpredictable, and random kinship patterns in which relatedness no longer flows teleologically (Franklin and McKinnon 2001) and thus do not represent the conventional modern "parameters of time and space" (Spielmann 1999, 140). Disrupting their respective kinship systems, digital and genetically engineered reproductive techniques blur the flow of conventional generations of images and bodies (see Butler 2000 and chapter 1).

If modernity's loss of aura is the "memory of religion in art," perhaps genetic engineering presents a loss of bio-aura as the memory of nature in biology (Baecker 2003, 13). The authenticity and stability of the image and of the body are thrown into question through new modes of reassemblage, recombination, substitution, and simulation. The convergence of digital and genetic manipulation produces a disturbance to existing modes of perception: if digitization potentially scrambles the spatiotemporalities of the visual image, geneticization puts the integrity of the reproductive body into question. Both digitization and geneticization have altered the authorial status of the generative impulse (in creating new forms of art and new forms of life) and raise questions about the ownership of the outcome (the new image or sound and the new cell, fetus, or life-form). If originals are at stake in the former (art as mimesis), then origins are centrally at stake in the latter (science as mimesis). The authentication of authorship and ownership in each case is established through shifting claims upon the successful and legitimate imitation of both nature and life.

If a living organism is a system that reacts independently to individual existence, then the dynamic image system that consists
of multi-sensory variables and reacts to input is also a living
organism.   WEIBEL, "Postontologische Kunst, 1994"

There is here [in cloning], perhaps, some obscure desire to obliterate the specificity of the species by genetic confusion . . . this
fated character of the Same . . . becomes materially inscribed
in our cells . . . science itself becomes a "fatal strategy." Singularity being, by definition, what can never be reproduced, all that
can be reproduced to infinity can only ever be of very low definition.   BAUDRILLARD, *Screened Out*

[8]   Leading Across the In-Between

Transductive Cinema in *Teknolust*

The reconfiguration of the human-technological interface cuts across
both of the above quotations, leading us to ask: wherein lies the specificity of organic life, and in what ways can it be imitated by technology? In Peter Weibel's conviction that a dynamic and responsive multi-
sensory image system could be understood as a living organism, we see
life defined as a reactive autonomous system. In Jean Baudrillard's declaration that, as a "fatal strategy," cloning enacts a deceitful imitation of
life, we see the particularity and singularity of human life destroyed by
the culmination of modern technological insanity (see chapter 1). Pulling in almost opposite directions (toward life's reiterative potentiality or
the end of life as we know it), these two quotations share a concern with
the transformation of the vitality of human life in relation to the artifice of digital and informational systems, on the one hand, and genetic
engineering and cloning, on the other hand. In chapter 3, I examined
how the genetically engineered reproductive body, imagined as informational and computational code, demands new forms of legibility and

intelligibility. In this chapter, I investigate the immersive and citational blending of cinematic, biogenetic, and informational technologies, and their respective reconfigurations of boundaries between the living and the nonliving, the organic and the artificial, and the animate and the inanimate. This chapter follows the previous one closely in its concern with disturbances to the previously foundational claims of both image and body presented by digital and genetic manipulation, and with the technical intersections of their shared loss of singularity.[1]

Opening with an animated strand of spiraling, fluorescent blue DNA which appears and disappears against a black background to the haunting rhythm of repetitive electronic music with a female vocalise, *Teknolust* (Lynn Hershman-Leeson, 2002) performs a series of cinematic citations to set the stage for what later is revealed to be a biomediated plot based on digital cloning. Two DNA strands in the same deep-blue tones frame the outline of a woman's face dappled by pixelation, as she gazes directly at the camera. As the same woman (Tilda Swinton) then showers and dresses for seduction in front of a triptych mirror like a 1940s femme fatale, the shot sequence moves from single to double to triple reflections, culminating in a shot of the woman standing in front of her multiple reflections in the divided mirror in front of her (see figure 30). Referencing a number of Hollywood's most famous noir films, such as *Double Indemnity* (Billy Wilder, 1944) and *The Lady from Shanghai* (Orson Welles, 1947), the scene demonstrates the long-standing association of femininity, image, and treachery about which feminist critics have written so eloquently.[2] As in classic film noir, the femme fatale's lethal desirability lies in her duplicity as enigmatic image emphasized through the use of diegetic frames within frames. As in the Hollywood originals in the 1940s, the masquerade of femininity is built into the cinematic apparatus itself.[3] The performative display of artifice transforms the double into the multiple and duplicity into multiplicity, prefiguring the narrative exploration of the biological imitations that follows. Exaggerating the clichéd iconography of feminine desirability (long, sleek hair; white skin; and matching red nail polish and lipstick), the opening scenes move from shots in science-fiction style of fluorescent blue DNA to a classic noir seduction scenario. The film's haunting, dreamy, electronic soundtrack, combined with these familiar elements of noir mise-en-scène, invoke both a nostalgic sense of cinematic fanta-

30. Ruby (Tilda Swinton) dresses as a femme fatale in front of a threefold mirror in *Teknolust* (Lynn Hershman-Leeson, 2002).

sies of idealized femininity and a feeling of anticipation of the futuristic artificial possibilities of science fiction. Sharing the deep blue of the spiraling DNA before her pixelated image is transformed into her "live" celluloid version in the shower scene, the femme fatale's presence here indicates the combination of the potential animations and mutations of science and the cinema.

Defying Fredric Jameson's description of pastiche as "blank parody, parody that has lost its humour" (Jameson 1983, 114), *Teknolust* is a witty pastiche of motifs from science fiction and film noir, blended with the tongue-in-cheek tracking of the mobility and mutability of informational matter that flows between human bodies and artificial life-forms, and between the cinema and computer screens.[4] Produced, written, and directed by the independent artist and filmmaker Lynn Hershman-Leeson, *Teknolust* synthesizes multiple incarnations of figures, scenarios, and interactions from her previous cyberworks, installations, and performances and her first feature film, *Conceiving Ada* (1997). Her interactive networked installation, "The Dollie Clones" (1995–98), is reworked here as the digital triplets who are the film's multiple protagonists. Citing the artist's repertoire of configurations, *Teknolust* is saturated with opportunities for a knowing audience to get the joke or recognize the reference to the world of art, science, experimental performance, and film (see Tromble 2005).

Richard Dyer (2007) argues that pastiche best describes a form with imitative intent which seeks to be appreciated as imitation, and *Teknolust* builds the pleasures of recognition into its formal artistic and cinematic references. Just as cycles of popular genre speak to the cinema audience's cumulative knowledge of the genre's histories, so this independent film engages in a dialogue with its audience by recycling past incarnations of the director's and other artists' work on the technologized and geneticized body. As the medium's own history becomes a resource for sustaining the three digital clones physically, citation becomes the central trope for playing with the interchangeability of biological and cultural modes of imitation. The mise-en-scène throughout the film is governed by a preoccupation with imitative repertoires, through which genetic engineering is embedded within the interplay of technological convergences. The increasingly co-constitutive potentialities of scientific and cinematic techniques are explored through the film's citational aesthetic. Permeated by references to the place of the figure of the woman within the history of cinema, *Teknolust* literalizes the notion of Hollywood as the dream machine, while pushing to tropic excess the place of the woman as object of Hollywood's structuring masculine desires. Bringing the much-debated scientific and cinematic drives to control the sexual and reproductive body of the woman within a shared frame,[5] the film foregrounds the processes of the artifice of femininity within each, using deadpan humor, absurd reiterations, and multiple intertextual citations. *Teknolust* is perhaps best characterized as art film pastiche that speaks back to the normative figurations of gender, sexuality, and reproduction of the mainstream of both science and the cinema.

The film's narrative concerns a biogeneticist, Rosetta Stone[6] (Tilda Swinton), who has secretly downloaded her own DNA and combined it with computer software to create three self-replicating automatons (SRAS), also all played by Tilda Swinton, who look human but function as intelligent machines. In order to survive, the SRAS need regular injections and infusions of sperm or semen to top up their constantly depleting levels of Y chromosome; thus Ruby, one of the SRAS, seduces random men to retrieve sperm for herself and her clone sisters. When a mysterious virus begins infecting men in the area, Dr. Stone is suspected of illegal genetic experimentation by her colleague Professor

Crick (John O'Keefe), and Security Agent Hopper (James Urbaniak) is called in to investigate her. Since the virus is infecting only men (causing impotence and a strange barcode to appear conspicuously on their foreheads), Dirty Dick (Karen Black), a "gender terrorism" expert whose nose for imposture seems to come from her own transgendered history, assists with the investigation. Meanwhile, Ruby develops a romantic attachment to her neighbor Sandy, a shy, boyish employee at the photocopy shop who is more concerned with conducting the rhythms of the machines than with the clarity of the copies they produce.[7] Making a pastiche of the gendered conventionality of film noir, science fiction, and romance, the increasingly absurd narrative moves toward closure through citational investigations of these genres.

The film's main preoccupation is not with solving the genetic crime (which in noirish style is inextricable from feminine sexuality), but rather with the convergence of the conventions of Hollywood cinema and scientific imaging in the context of the new imagined vitalities of digital and cloning cultures. Defying expectations generated by the promise of the film's title (and by its pseudo-erotic opening), the sexual seduction performed here is marked by its brevity and almost scientific efficiency, rather than by passion or eroticism. A series of close-up shots fading to black appear in rapid succession, accompanied by the rise and fall of deep breathing which gives a biological, sexual rhythm to the visual sequence. The woman appears in control of what seems to be a repetition of a highly rehearsed sequence, the sense of precision in this bizarrely ritualized sexual scenario giving the spectator a series of clues. Like the conventions of the genres it cites, the very tight set of rules of the mise-en-scène of the opening seduction connect the artifice of science and the cinema through a geneticized sense of rehearsal, repetition, sequence, and predictability.

The three digital clones embody a tension between sameness and difference which makes their hidden (and clandestine) imitative origins visible on the surface. Half-human, half-machine, these recombinant triplets are an amalgam of genetic and computational material, embodying what Adrian Mackenzie calls the "indeterminacies of biotechnology" (Mackenzie 2002, 176). They are both living and nonliving, human and nonhuman, authentic and artificial, individual and multiple; they are Rosetta Stone's cyborg offspring, both her sisters and her daughters, both like and unlike their progenitor and each other. The sense of their artificial genetic sameness is humorously produced through a rhymed visual aesthetic of shape, surface, texture, and contour (including their identical silk kimonos and the matching bed linen and minimalist décor in their rooms), but their different colors denote a type of comic-book individuality: Ruby is the red one, Marinne the blue one, and Olive the green one. The mimicry of the logic of the color coding of the RGB (red, green, and blue) components in video computer signals refers to the technological convergence between digital and genetic sampling. Shot with a 24-frame, high-definition digital camera and later transferred to film, the vivid, largely primary-color aesthetic of *Teknolust* has a comic-book sensibility, giving it a hyperreal quality in which the boundaries between the real and the virtual become increasingly confused.

This difference in sameness is articulated through an almost cartoonish color coding of the distinctions in the clones' physical appearance: the brunette, the blonde, and the redhead—one of each type of traditional white (but colored) femininity. If the different colors of the clones' hair, clothes, and bedroom décor offer a visual pun on the way Hollywood has circulated different types of white femininity (ultimately variations on the same fantasy), then the artifice of difference that is made visible through these conventionalized Western ideals is also inflected with an Easternized aesthetic. The geisha-style sleek black hair, pale white skin, and silk kimono, together with the ritualized precision of semen collection and infusion preparation, give Ruby's service to her siblings an almost Orientalized quality (see two images in figure 31).[8] The play here on using color as an index of difference highlights the question of how shared biological origins translate into visible signs: how can we overcome the problem of otherness if the same threatens to be literalized as the identical? Anxieties about the legibility of difference on the surface of the body when faced with replications and multiplicities (in both the racialized and the geneticized senses) are resolved for the spectator through the literalization of the idea of color coding (if you can't see the difference, the color coding will guide you). As the problem of the genetically identical being physically indistinguishable is worked out through color coding, these visible differences among the white clones transform biological distinctions into cultural ones, or rather articulate genetics as aesthetics.

Digital cloning is given a visual life through the presence of Tilda Swinton simultaneously playing the three clones and the scientist in different combinations on screen, achieved either with the use of

31. Ruby's (Tilda Swinton) (1) long, glossy black hair and pale skin and (2) acts of service performed in her silk kimono give her femininity a "geisha" look in *Teknolust* (Lynn Hershman-Leeson, 2002).

32. The three digital clones, Ruby, Marinne, and Olive (all
played by Tilda Swinton) appear together in the same room
in *Teknolust* (Lynn Hershman-Leeson, 2002).

doubles or with a split screen and digital composition (see figure 32).
Her characters' lack of physical resemblance to each other (and to the
scientist Rosette Stone who invented the three cloned offspring with
her DNA) produces a fascination with her facial features, expressions,
and gestures, making the spectator want to look beyond the surface
differences of hairstyles and kimono colors to see the visual evidence of
her performance. An actor whose ability to not quite fully inhabit her
own body is strikingly suited to this casting of her as original and multiple copies. Swinton achieves a sense of embodiment in between the
human and nonhuman, on both sides of the biological-technological
divide.

Transduction in microbiology is defined as "the transfer of genetic material from one cell to another by a virus or virus-like particle" (OED).
In *Transductions: Bodies and Machines at Speed*, Mackenzie argues
that biotechnology actually "heightens the experience of complex inseparability between the living and the non-living" (Mackenzie 2002,
175). He suggests that when we think about how to formulate what is at
stake in biotechnology, we need to "highlight a formative and radically
contingent *collective* entwining between living bodies and information,

rather than an interfacing of individual bodies with machines" (ibid., 172, emphasis in the original). Instead of assuming that something "living animates technology," Mackenzie contends that biotechnology involves "a kind of design, and a kind of engineering, but a designing that intimately associates living and non-living elements," producing a troubling status which challenges traditional systems of representation and thought (ibid., 175–76).

Following the French philosopher Gilbert Simondon, Mackenzie attempts to redress what he considers the "misguided opposition between culture and technology" inherent in notions such as the human-machine interface (ibid., 11). Pushing beyond traditional conceptualizations, he examines how biogenetic technologies (or "technicities," as he calls them) reconfigure the relationship between human bodies and forms of technical artifice in the current "distributed ensemble of living and non-living actors" that constitutes biogenetic technology. What he calls the "*indeterminate status* of this complication of the living and the non-living" captures the problem of how to imagine the oneness of the two (or even the three) of cloning (ibid., 175–76, emphasis in the original). For Mackenzie, technicity "refers to a side of collectives which is not fully lived, represented or symbolized, yet which remains fundamental to their grounding, their situation and the constitution of their limits" (ibid., 11).

Mackenzie believes the concept of "transduction" offers a potential alternative to the limits of current thinking about the misconceived "human/non-human interface" of biotechnology, for it helps reframe the "thinkability" of technology, highlighting a "margin of contingency" or indeterminacy which "participates in the constitution of collectives" (ibid., 3). As he explains, "the hallmark of a transductive process is the intersection and knotting together of diverse realities" (ibid., 13). Mackenzie ties together technicity and transduction by arguing that technicity is a kind of transduction, since technicity, to use Simondon's words, is that "quality of an element by which what has been acquired in a technical ensemble expresses and conserves itself in being transported to a new period" (Simondon quoted in ibid.). According to Simondon's definition of transduction:

> This term denotes a process—be it physical, biological, mental or social— in which an activity gradually sets itself in motion, propagating within a given domain, by basing this propagation on a structuration carried out

in different zones of the domain: each region of the constituted structure serves as a constituting principle for the following one, so much so that a modification progressively extends itself at the same time as this structuring operation . . . The transductive operation is an *individuation in progress*; it can physically occur most simply in the form of progressive iteration. However, in more complex domains, such as the domains of vital metastability or psychic problematics, it can move forward with a constantly variable step, and expand in a heterogeneous field. (Simondon quoted in ibid., 16, emphasis added)

For Mackenzie, Simondon's emphasis on ontogenesis (how something comes to be) rather than ontology (what something is) unites the concepts of both transduction and technicity and marks "a mode of thought focused on a unity of becoming rather than a unity of substance" (ibid., 17). In molecular biology, transduction is the name of the process by which a virus carries genetic material into the DNA of a bacteria, but examples of transduction occur in the physical realm (for instance, the growth of a crystal suspended in liquid) and in the mental processes of thought and affect where "constantly varying rhythms" oscillate in a "field structured by differences and repetitions" (ibid.). Transduction importantly arises from the "non-simultaneity . . . of a domain"—or, to put it another way, the fact that the domain is "not fully simultaneous or coincident with itself" (ibid.).

The original meaning of "transduce" was to lead over or to move from one place to another; the word comes from the Latin *trans* (across) and *ducere* (to lead). While the consistent element of the word's diverse definitions is transformation, the focus of the transfer varies. In the twentieth century, the verb was used in a number of technical contexts to mean "to alter the physical nature or medium of (a signal); to convert variations in (a medium) into corresponding variations in another medium" (*OED*, definition 1). A transducer is "any device by which variations in one physical quantity (e.g. pressure, brightness) are quantitatively converted into variations in another (e.g. voltage, position)."[9] What is interesting about the etymology of the term is the variability of the transpositions at stake: in acoustic measurement, a microphone translates a message; in neuroscience, the rods in the retina transform light; in microbiology, genetic material is constant but is transported from one site to another; and sound waves are transformed into electric currents, while the variation is constant.

Transduction thus refers to multiple conversions, alterations, and transportations of matter, medium, and pattern, for it "aids in tracking processes that come into being at the intersection of diverse realities [which] include corporeal, geographical, economic, conceptual, biopolitical, geopolitical and affective dimensions" (ibid., 18). If transduction entails a "knotting together of commodities, signs, diagrams, stories, practices, concepts, human and non-human bodies, images and place" which involve "new capacities, relations and practices whose advent is not always easy to recognize" (ibid., 18), then for the purposes of this chapter, transduction might provide a conceptual route through the complexity of new life-forms in the convergence of genetics, digitization, and the cinema. How might transduction help us think through those phenomena whose "elements are assembled from non-living and living milieus" (ibid., 23)? We might place the "distributed ensemble of living and non-living actors" of biotechnology in dialogue with the distributed ensemble of living and nonliving actors on the digital cinema screen, in the interplay of the illusory and the real, the absent and the present, and the live performance to camera and the virtual. We might ask how transduction helps us think about the ways in which the living and the nonliving are "grafted onto one another" cinematically (ibid., 204).

For the purposes of examining the sexual charge of particular modes of transduction, we might also trace the word back to its etymological base: traduce. This word has a much longer history than transduce, dating back to 1535. It also means to lead or bring across, but it has a number of other meanings significant to the current analysis: traduce (from the Latin *traducere*) "to lead across . . . ; also, to lead along as a spectacle, to bring into disgrace . . . to speak evil of, esp. (now always) falsely or maliciously; to defame, malign, vilify, slander, calumniate, misrepresent; to blame, censure . . . to expose (to contempt); to bring dishonour upon, dishonour, disgrace . . . to lead astray, mislead, seduce, betray" (OED, definition 3a, 3c, 4). This older usage speaks to questions of desire which are absent from its more recent iterations and from Mackenzie's discussion of them, but which are central to my consideration of fantasies of biotechnologies here.

Teknolust presents Ruby's body as the site of convergence of two imitative modes. As an electronic clone, she already embodies the grafting

33. Ruby (Tilda Swinton) sleeps, as a romantic scene with
Elizabeth Taylor and Van Johnson from *The Last Time I Saw Paris*
(Richard Brooks, 1954) is projected over her body in *Teknolust*
(Lynn Hershman-Leeson, 2002).

of the living and the nonliving onto one another, and her recombinant
materialization is performed throughout the film in the scenes of her
acquisition of the art of feminine seduction from female stars of Hollywood cinema—specifically, Hedy Lamar, Elizabeth Taylor, and Kim
Novak. The projection of classic romantic scenes from 1940s films across
Ruby while she sleeps is Rosetta's solution to the problem of Ruby's lack
of an inherent (or inherited) capacity for seduction, due to her absence of human drives (see figure 33). Following her nighttime absorp-
tion of data on heterosexuality, Ruby randomly selects male strangers
as partners (any man will do), seducing them with the decontextualized
lines of romantic heroines in the films, such as: "don't let the celebration ever end" (Helen Ellswirth (Elizabeth Taylor) in *The Last Time I
Saw Paris* [Richard Brooks, 1954]) and "you're looking good tonight,
Frankie, you've got natural rhythm" (Molly [Kim Novak] in *The Man
with the Golden Arm* [Otto Preminger, 1955]). These surreal exchanges
turn heterosexual seduction into a bizarre repetition of a genre, drawing attention to the banality of its content and the predictability of
the male heterosexual response (the men would all be willing sexual

partners even if Ruby whispered the Greek alphabet to them). The film plays sexual stereotypes and cinematic clichés back to its audience in a deadpan style. The human and the nonhuman become almost indistinguishable here, as the recharging of what Marsha Kinder (2005, 169) calls "the intelligent machine" through sleep blends with the recharging of heterosexuality through cinematic and computational input. The cinema as a technology of idealized feminine heterosexuality is taken to comic absurdity as Ruby's cloned body becomes the site of this doubled mimicry.

Writing of the creation of intelligent machines, Baudrillard suggests:

> If men create intelligent machines, or fantasize about them, it is either because they secretly despair of their own intelligence or because they are in danger of succumbing to the weight of a monstrous and useless intelligence which they seek to exorcise by transferring it to machines, where they can play with it and make fun of it. . . .
>
> If men dream of machines that are unique, that are endowed with genius, it is because they despair of their own uniqueness, or because they prefer to do without it—to enjoy it by proxy, so to speak, thanks to machines. (Baudrillard 1993, 51)

Following Kinder's reading of *Teknolust* as a feminist intervention in the patriarchal relations of scientific knowledge production, we might respond to Baudrillard by asking: what if *women* invent intelligent machines or dream of machines that are unique? Kinder reads *Teknolust* as part of the director's contribution to "a rich tradition of feminist works on technology and the body" (Kinder 2005, 178). Celebrating the challenge to the masculinity of scientific genius, Kinder suggests that, like *Conceiving Ada*, *Teknolust* "really focuses on the relations between two equally brilliant women (here Ruby and Rosetta) who are both seeking agency in different realms and media and whose shadow relationship helps empower not only both of them but also the rest of us" (ibid., 180). In many ways, she suggests the film belongs to a mischievous and celebratory strand within feminism that began with Haraway's "A Manifesto for Cyborgs" (1991) and that has the posthuman offering ways of reimagining the embodiment of gender, sexuality, and reproduction (ibid., 178).

As Kinder argues, the concept of downloading is key to *Teknolust*: Rosetta's "nightly ritual of downloading old Hollywood movies to prepare Ruby for 'real-life' encounters . . . evoke[s] Hershman's own intertextual strategies, which try to reprogramme those of us who watch her movies" (ibid., 178). The reiterative imperatives of heterosexuality as the site of conventionalized sexual relations extend from the re-screening of downloaded Hollywood films to Ruby's own reproductive quest. Biological formulas blend with cultural iconographies in a mise-en-scène of prosthetic mutability and interchangeability: the monitors in the clones' bedrooms function as both cinema screens and modes of surveillance (see figure 34); the microwave in the kitchen doubles as a communication system (see figure 35); the photocopy machines in Sandy's shop together produce orchestral sounds as they print. Just as the cinema becomes data for prosthetic sexual seductions, so these in turn produce biological components necessary for the sustenance of artificial life. Like vampires or addicts, the clones must have their daily shots and brews, and Ruby will go to any lengths to get their required male substance. The distinctions between human and nonhuman materialities dissolve as media messages program the sexual behavior necessary for the chromosomal input required for the electronic clones' survival. As cellular appetite replaces sexual appetite, heterosexual seduction becomes a biological requirement for the survival of this new hybrid species.

In this deception of her male victims in the service of the reproduction of artificial life, Ruby becomes the transducer insofar as a "transducer" is "any device by which variations in one physical quantity (e.g. pressure, brightness) are quantitatively converted into variations in another (e.g. voltage, position)." In *Teknolust* Ruby secures chromosomal variation in one physical quantity (semen) in order to convert it into another (the life blood of artificial cloned bodies). To achieve this conversion, Ruby's body performs another, absorbing celluloid seduction scenes and reenacting them to acquire the requisite biological sustenance. As the transducer, Ruby's body is the switching point between cloned artificiality and biological dependence, and between celluloid fantasy and male heterosexual desire. Defined as a "progressive iteration" at its most basic level, transduction perfectly defines the mode of Ruby's seductions through her performance of Hollywood femininities, which result in the conversion of one substance into another (semen into artificial life).

Transductions between physical, imaginative, and creative domains

34. Rosetta (Tilda Swinton) gives parental instruction on the screen in Marinne's (also Tilda Swinton) room in *Teknolust* (Lynn Hershman-Leeson, 2002).

35. Ruby (Tilda Swinton) communicates to Rosetta (Tilda Swinton) through the kitchen microwave in *Teknolust* (Lynn Hershman-Leeson, 2002).

proliferate as the film plays with the indeterminate status of bioinformatic material. Ruby's role as transducer extends to include the conversion and transmission of a virus from a computational to a biological form. The computer virus downloaded by mistake into Ruby's body at her conception becomes a biological virus infecting all her male victims, who manifest identical physical symptoms: impotence, lack of appetite, and the appearance of a barcode on the forehead. In the scene in which Ruby's infected sexual partners have been isolated for medical observation (under a national security alert as victims of gender terrorism), their uniform appearance in hospital gowns, bagged feet, and surgical caps offers a ridiculous vision of their shared status as objects of the medical gaze. In contrast to the individualized color-coded clones, these men become a uniform, undifferentiated mass, shuffling around the room in a group, peering at us through the lens of the surveillance camera. They finally express their shared status by huddling together in the shape of an X, giving comic chromosomal form to their collective sexual status as unwitting sperm donors—male victims of an insatiable female appetite (see figure 36). Their place as sexual and reproductive commodities serves as comic punishment for their genital predictability. Ruby's calculated seduction of her male prey for the purposes of extracting their sperm (rather than for money or power, in the usual noirish tradition) thus plays the normative masculine heterosexual drive (for which any woman will do) back upon its own conventional instrumentality.

As the physical conductor responsible for the men's infection, Ruby the femme fatale brings sexual deception and transduction together. Following the associations of the older meanings of "traduce" (to dishonor, disgrace, mislead, seduce, or betray), this connection deepens: the femme fatale's manipulative seductions in *Teknolust* lead men across the threshold of their desires into a fantasy world of Hollywood reenactments and, in so doing, expose the reiterative clichés of their predictable sexual drives. Susceptible to the appearance of a digitally or genetically engineered ideal, the men, whose samples of semen line Rosetta's kitchen cupboard in a row of jars (like notches on a bedpost), labeled by name and mug shot, are victims of their own programmed desires which Ruby manipulates so successfully. Her transductive performances, through which the men are seduced by lines from old Hollywood movies, mock the predictability of their sexual response through

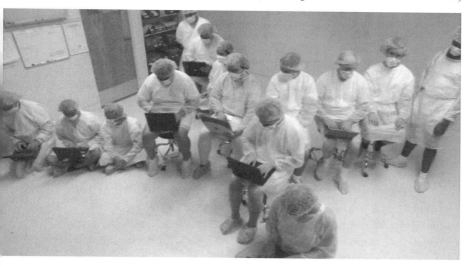

36. Men in isolation because of the virus Ruby has given them form an X shape in *Teknolust* (Lynn Hershman-Leeson, 2002).

random repetition. With each new seduction, the conventionality of heterosexuality is reiterated and its absurd logic exposed.

Thus, Ruby's power as seducer depends upon her role as transducer. Her imitative deceptions disturb the predictable continuity across the shifting ground of these transductions (and their transcontextual mobility). If the normativities of the technical element of transduction is defined by the stable configuration that "expresses and conserves itself in being transported to a new period," then these rehearsed seductions reveal the men as ultimately sharing the same sexual response (Simondon quoted in Mackenzie 2002, 13). Ruby's nighttime rituals work here as transductive enactments which ridicule the normative repetitions of "compulsory heterosexuality" (Rich 1980). In converting celluloid to real life, and biological substance to artificial life, Ruby simultaneously demonstrates the mobility of matter and reveals the enduring cliché of transcontextual heterosexual masculinity.

Like a number of films discussed in this book, *Teknolust* plays these questions of artificiality and imitation across the heterosexual-homosexual distinction (see chapters 2 and 5). Ruby's vampish performance of heterosexual femininity is matched to Olive and Marinne's same-sex, incestuous coupling. As Rosetta says, "Ruby has to seek out intimacy, but Olive and Marinne have each other." The triangular dynamic of the clone triplets operates across a distinction between Ruby as the singular femme fatale searching for heterosexual prey (in need of a counterpart, only half of a gendered whole, one of a pair, and so on) and Olive and Marinne as the cohabiting lesbian-continuum cloned couple who remain firmly within the zone of the artificial in sexual and reproductive matters (see figure 37). As Ruby's femme fatale dutifully transforms sexuality into reproduction, the lesbian duo embodies a visualization of genetic artifice in their color-coded intimacies. Ruby's dependent singularity demonstrates the heteronormative imperative of the clones' design: downloaded from only female DNA, they are all condemned to an eternal dependence upon male chromosomal input (especially when Rosetta's artificially engineered semen ceases to work). Heterosexual complementarity is inscribed here within the biological trajectory of their survival: male DNA (via sperm) is their life source. If Ruby's vampish seductions contrast with Olive and Marinne's decidedly domesticated and unerotic relationship, Ruby's own lack of erotic motivation in her utterly instrumental approach to sex aligns her with her siblings: in the end, all Ruby really wants to do after obtaining her samples of ejaculate is to cuddle with her male victims. Refusing the spectator the pleasure of erotic identification, the film mocks the voyeuristic expectations its title evokes. More *Teknocuddle* than *Teknolust*, the sense of sexual intensity promised by the film's title is only to be found in the desire for the technological that it aestheticizes.

Sexual artifice is elaborated further through the figure of Dirty Dick (a female private detective who used to be a man). Famous for her performance as the transsexual character (Joanne) in *Come Back to the Five and Dime Jimmy Dean, Jimmy Dean* (Robert Altman, 1982), Karen Black appears here as a worldly and intuitive (trans)gender consultant, echoing the cyborgian imitations of the clones with her own multiple imitative embodiments, secret past, and queer origin story. A woman playing the part of a male-to-female transsexual, Black performs multiple masquerades. As a private detective with a nose for the truth about people, she reads the affect of others correctly, and it is her intuition and empathy which eventually solves the mysteries of the cloning and the seductions, and elicits Rosetta's "talking cure" confession. Dick becomes Ruby's counterpart: not the transducer of erotic currents, but the conduit for human affect despite the artifice of the world (and the body) she inhabits.

37. Olive and Marinne (both played by Tilda Swinton) embrace in *Teknolust* (Lynn Hershman-Leeson, 2002).

The mysterious gender terrorism that Dirty Dick is called in to solve is an unforeseen consequence of Ruby's conversions. In her inadvertent transmission of a computer virus to the electronic clones, Rosetta traduces an informational error into a biological one, while transducing one form of materiality to another. (One further definition of traduce is "to pass on to offspring, or to posterity; to transmit, esp. by generation"; OED, definiton 2). When Ruby continues the viral legacy in her sexual transmission to her multiple male partners, the error is transduced into human form. This association of transduction with transmission through kinship and generation moves Ruby's seductions in *Teknolust* in a diachronic frame, since they both result from a transmission by a previous generation (the borrowed lines from 1940s stars) and enable a new generation to thrive (the clones as daughters of Rosetta). Insofar as transduction is the "quality of an element by which what has been acquired in a technical ensemble expresses and conserves itself in being transported to a new period" (Simondon quoted in Mackenzie 2002, 13), Ruby's cloned body becomes the transducer of female desire from one artificial mode to another. The cure to the infection stretches this play on artifice a step further in Olive's and Marinne's bid to generate an offspring through technical means. Like a technologized primal scene, they rebel against Rosetta's prohibitions and break into the mainframe, finding the secret formula for their own technical origins and using it to engineer a new clone for themselves. In the process, they discover the computer virus infecting Ruby's victims and, in eradicating it on Rosetta's laptop, simultaneously cure the male patients in hospital of their (hysterical) symptoms. A miracle healing, the virus is purged across these diverse materialities (significantly without any direct physical contact). Reversing the original heterosexual infection, the lesbian couple produces the ultimate transductive fantasy of prosthetic materialization from informational to biological bodies.

These different technical flows of information converge in the body of Ruby as transducer. The absorbent and mediating capacities of her body generate new forms of relatedness between different kinds of materiality. Through these kinship innovations, Ruby's body condenses biological, electronic, and celluloid sexual and reproductive femininities. Hollywood cinema has been seen as the twentieth-century dream machine, circulating fantasies of female sexuality that fulfilled the needs of the male psyche,[10] but here its function is reversed, as it feeds the genetic needs of the female clones in a new circuit of exchange: the celluloid image becomes sexual data, securing biological substance to sustain the lives of the clones. In the scenes of downloaded celluloid sampling, film clips of classic Hollywood stars (what Rosetta calls a "motivational tape") are projected across the sleeping Ruby, her prosthetic body becoming a cinema screen. The projected images appear simultaneously both on her and on the large computer screen on her bedroom wall, flickering across her body, especially her face, and placing one imitation of life in literal physical contact with another—the sight of them touching is the sign of the successful programming. The images of the film stars are life-size or bigger, as these two forms of feminine artifice meet in an intimate ritual of fusion.

In the most extended of these recharging scenes, the digitized, highly pixelated film clips of Frank Sinatra and Kim Novak in *The Man with the Golden Arm* are shown first on the cinema-sized computer screen, then projected across the sleeping Ruby, transforming the celluloid image into data and producing its illusion of realness as an imitative technical effect. The cinematic image is literalized here, as the artifice of the 1940s Hollywood stars gives visual form to Ruby's own artificial origins. At one point in the scene, the composition of the shot places her body between those of the two stars in their romantic encounter; this diegetic proximity offers a virtual caress between the image of the celluloid stars and the electronic clone. This tongue-in-cheek cyborgian threesome is one of the many sexual triangles in the film which play with the cinema's Oedipal reputation. Ruby (Swinton) then turns her face to meet that of Molly (Novak): nose to nose, lips to lips, in an erotic rhyming of multilayered artificial female forms (see figure 38). When Ruby's face is matched with the close-up shots of Molly's, this proximity anticipates the ventriloquism of the following night, when the former recites the words of the latter to her next victim. The obvious intertextual reference here is to Novak's famous performance as Judy/Madeleine in *Vertigo* (Alfred Hitchcock, 1958), an actress imitating a male fantasy of feminine perfection for which Scottie (James Stewart) naively and predictably falls. The cinematic image moves across Ruby's sleeping body like a projection of her dreams, as the sound of Molly's line—"you're looking good, Frankie, you've got natural rhythm"—repeatedly echoes with almost ghostly tones. With the distortion of both sound and image, the projected film clip becomes mobile information, moving between the past and the present like a prosthetic memory haunting Ruby's dream world. The psychic and the cinematic are tied together in these multiple projections that transform the apparently previously stable phenomena of celluloid image and biological body into new forms of mutating informational flow.

The sensory connection between the cinematic image of a woman and her contemporary cyborg counterpart enacts what Mackenzie calls their ontogenesis—their shared becoming (Mackenzie 2002, 17). Both produced by techniques of desire (one a screen fantasy, the other cloned automatons) their nighttime touch transduces femininity across their common artificial materialities, uniting them through intimacy. Baudrillard suggests that the "computer generated voice isn't exactly a voice" and "looking at a screen isn't exactly looking," and that "the tactility here is not the organic sense of touch: it implies an epidermal contiguity of eye and image, the collapse of the aesthetic distance involved in looking." In our increasing proximity to technology, he argues, "we draw ever closer to the surface of the screen; our gaze, as it were, strewn across the image" (Baudrillard 1993, 55). In this scene, the image is strewn across our gaze and indeed is strewn across Ruby's body. *Teknolust* plays with precisely such uncanny confusions of organic and inorganic categories through the biologically and culturally

38. Ruby (Tilda Swinton) in virtual intimacy with the projected image of
Molly (Kim Novak) in *The Man with the Golden Arm* (Otto Preminger,
1955) in *Teknolust* (Lynn Hershman-Leeson, 2002).

reproductive body of the woman whose status has come to signify the
long-standing problems of such distinctions.

These transmissions of femininity signal the convergence of celluloid
and digital techniques. In copying Hollywood heroines with computational accuracy, Ruby performs an exaggerated version of the dynamic
between women in cinema audiences and Hollywood stars as their ego
ideals.[11] The performances of romantic love by Hollywood's heroines
on screen become a digital resource for an instrumentalized feminine
heterosexuality—a means to an end to be sampled, cited, and recycled.
In Ruby's imitations of heterosexual seduction in the service of the imitation of life itself, even desire becomes information. Heterosexuality is
staged as instrumental copies of a copy for which, as Judith Butler has
famously argued, there is no original: "as imitations which effectively
displace the meaning of the original, they imitate the myth of originality
itself" (Butler 1990, 176). A literal dream machine, Hollywood cinema
in *Teknolust* becomes the key component in a transductive relay of imitations: its scenarios of romance and seduction are reiterated as generic
performatives, whose enactment guarantees the desired effect on the
male recipient (with one notable exception). Geneticized heterosexual-

ity here enacts a performative Hollywood citation. As data used to secure sperm, Hollywood films are a vital life source for the clones; as an
imitation of life on the screen necessary for the imitation of life beyond
it, the cinema services science. Or rather, the two modes of imitation
blend in a networked circuit of exchange of virtual life-forms.

In these scenes of celluloid programming, we move beyond analogy
and the interface into the terrain of transduction. The parallels between
specific techniques involved in the cinema and cloning here are only
the starting point for a more profound set of materializations of biotechnical transformation. *Teknolust* animates techniques that perform
mutability in between the living and the nonliving, through which new
vitalities move across different material strata (Hayles 1999). Here the
film speaks to Eugene Thacker's question: "what would it mean to approach the body as media in itself?" (Thacker 2004, 9). Ruby's body
becomes a form of what he calls "biomedia": not the biological body
hybridized with or supplanted by machines, but the "intersection between genetic and computer 'codes' facilitat[ing] a qualitatively different notion of the biological body." Here, the "'bio' is transformatively
mediated by the 'tech', so that the 'bio' reemerges more fully biological"
(Thacker 2004, 6).

For Thacker, the convergence of information and genetic science re-
biologizes the body, recontextualizing it as a "body more than a body,"
where "the body you get back is not the body you began with" (ibid., 6).
As Ruby's cloned body enters these circuits of transmission and conversion, it generates the biomediations of a culture which wants to "render
the body immediate, while also multiplying our capacity to technically
control the body" (ibid., 9).[12] *Teknolust* gives cinematic life to the invisible gene through the desire for immediacy in the circuits of artificial
exchange and biomediation. Beth Coleman (2006) argues that multi-
media art concerns an aesthetic of the generative; *Teknolust* explores
such an aesthetic through its reconfiguration of bodily materiality and
its generative potentialities. It imagines the immediacy of becoming,
the immediacy of the generative.

The generativity of imitative kinship extends from bodies to images
in *Teknolust*. Ruby's citational relationship to Hollywood cinema is an
embodiment of the whole film's imitative borrowings from its more
mainstream predecessors, which becomes a technique for exploring

the new kinship of digital cloning in the film. Put simply, citation becomes a form of kinship exchange. *Teknolust* reconfigures cinematic
genealogy through an intertextual pastiche, pushing the conventions
of the genre to absurd limits and implicating our recognition of and attachment to these famous cinematic gestures and formulas. This sense
of critical, yet inevitable, indebtedness to previous iconographic codes
recontextualizes genre as a way of playing with the changing status of
the image more generally.[13] For example, the replay of the noirish duplicity of the femme fatale, suggested by the reflection in the mirror
shot, extends the problem of authenticity beyond the woman on the
screen to the medium of film itself (see figure 30). If film noir has cast
suspicion on the seductive woman through a mise-en-scène of shadows
and occlusions, her conventional surface deception here is rehearsed
as a deceit about the status of the image within the film's own diegesis.
The triple image of Ruby in the mirror cites the iconography of her
predecessors, while simultaneously visualizing her own duplicated and
genetically engineered kinship (see Franklin and McKinnon 2001 and
Butler 2000). Duplicity becomes multiplicity, as the femme fatale is revealed to be a cloned triplet.

Teknolust's citation of the centrality of woman as image (or as idealized spectacle) to the visual pleasures of film extends beyond the genre
of noir into the history of Hollywood cinema more generally. In one
scene, all three SRAS appear on screen simultaneously, dressed in identical yellow silk kimonos (the only primary color not already claimed by
the RGB clones) and performing the dance sequence they have invented
for Rosetta when she returns from work. Like a number from a Busby
Berkeley film, the scene uses bird's-eye shots to capture the formal arrangement from above, mocking the pleasures of visual symmetry and
sequence found in the formal arrangement of female bodies through its
absurdly improvised aesthetic. This rather unimpressive, chaotic ensemble of random jerky movements by the triplets also rehearses something else: the technical ingenuity involved in the simultaneous screen
presence of Swinton as all three clones, bringing genetically engineered
reproduction and digital imaging techniques into a shared frame of deceptive ingenuity. Swinton's reputation as a performance artist, associated with experimental work, contributes to our sense of her suitability
as embodiment of both kinds of artifice.

Teknolust thus aestheticizes the imitative desires of genetic science
it narrativizes, delighting in using the history and techniques of im-

age cultures as an aesthetic resource for imagining a cloned future. The multiple artistic and cinematic citations offer the pleasures of visual recognition and repetition. The point here is not so much accuracy (few viewers will have the cultural capital to identify all the intertextual gestures) as it is the citational aesthetic that saturates the authentic with the imitative, and the real with the virtual. As the spectator is pulled simultaneously into the past (classic Hollywood) and the future (cloning), the film plays with the sedimented pluralities of inheritance and relatedness. Just as the SRAS combine living and nonliving elements, the whole film recycles cinematic styles from the past with live animations in the present.[14] The proliferation of citations connects the digital to the genetic through the idea of sampling and recombination. As downloaded DNA recombined with computer software, the SRAS embody the melding of biological and digital techniques, performed within a mise-en-scène of mimicry which recycles styles, images, texts, spoken lines, objects, and gestures. Technical mutability and fluidity blur the boundaries around these categories, connecting art, science, and the cinema in the current climate of commodified sampling and recontextualization.

The shifting significance of this living and nonliving ensemble is integral to *Teknolust*'s modes of gendered spectatorship, which slide seamlessly between live and virtual exchanges in a series of multiple screen relations. Ruby's e-dream portal appears both within the diegesis (see figure 39) and exists extra-diegetically in a continuing live form as a website.[15] In the film, Ruby's portal is a sexual interface offering intimacy to strangers so they can escape the loneliness of real life through a virtual and vital connection.[16] Like the heroines of Hollywood before her (whose scenes become her dreams), Ruby appears at her e-portal as a fantasy figure enticing visitors to imaginary places, as her electronic voice-over repeats: "I can teach you to dream—elope from your remote—evolve with me—let's dream together." The fantasies at the e-dream portal share the oneiric mood of Ruby's off-screen vampish seduction scenes. Her incarnations as Agent Ruby transform her from the passive absorber of Hollywood romance to the active generator of dreams for lonely late-night Web surfers. In both cases, male projections lie at the heart of the mediated encounter. Sandy's attachment to Agent Ruby on his laptop is no less affective through its mediated framing—his clandestine connections provide an immediacy absent in the rest of his surreal life. On both the cinema and the computer screen,

39. Ruby's e-dream portal in *Teknolust*
(Lynn Hershman-Leeson, 2002).

Ruby inhabits the illusory space of misrecognition—as human in her live sexual encounters, or as artificial persona in a virtual erotic fantasy as Agent Ruby online. Either way, Ruby is always passing.

Similarly, communication between Rosetta and the SRAS depends upon multiple mediations, constructing spectatorship within a relay of screens. Like a kind of virtual panoptic gaze, Rosetta's technological presence in the clones' world through her digital circuits allows close monitoring of her offspring. On Rosetta's home computer, each clone has an iconic presence as a color-coded double helix, through which Rosetta can also access details about their physical well-being: information here is the materialization of the genetically engineered body; matter appears mutable. Rosetta monitors the clones' Y-chromosome levels on her laptop "spermometer." Inscribing spectatorship in the mobility across different live and virtual interactions, shot sequences cut between the points of view of human and machine. In the spaces between the living and the nonliving in these biomediated animations, the film thus constructs a transductive subject leading across the in-between.

Overlaid with excessive conceptual punning, *Teknolust* pushes the idea of the reproducibility of reproduction in on-screen and off-screen artificial life to its associative limits (see Franklin and Ragoné 1997 and Roof 1996b). Writing of the convergence between two forms of artificial reproduction—the photocopy and the clone—Baudrillard calls cloning "a process of metastasis that began with industrial products [and] has ended up in the organisation of cells" (Baudrillard 1993, 120). For him, "the genetic code, substituting for the father and the mother, becomes the true universal matrix, the individual being now only the cancerous metastasis of his basic formula," resulting in "the displacement of heterosexual reproduction as the embodiment of difference necessary to the human psyche" (ibid., 199–200). Contra Baudrillard's vision, the bizarre romance between Ruby and Sandy in *Teknolust* delivers affective connection through their shared love of artifice and their sensitivity to the inauthentic. Their exchanges enclose them in a series of transductive circuits: Ruby's cyborg body absorbs Hollywood dreams when she sleeps, and her Web portal transmits e-dreams to surfers like Sandy who can't sleep at night. The film sutures the live and the virtual in a dialogic, autotelic exchange. Agent Ruby combines digital and genetic artifice when her facial features appear as a pixelated collage with a cellular feel. Like the flow between the live and the celluloid in the screening of Hollywood stars across Ruby's body, the scenes of the e-dream portal place her in the in-between space of screened vitality and live digitality.

Both Ruby and Sandy embody an affective relationship to artificial reproduction (digital and genetic, mechanical and celluloid). Their status as outsiders to conventions of live human interaction places them together in the romance of a shared space of mutual technophilic fascination. Sandy connects with Ruby through their preference for nonhuman forms of life and an appreciation of the culture of the copy: her lack of knowledge of human life matches his own lack of interest and confidence in it. Their low-key romance is sparked by his fascination with nonhuman reproduction (a photocopy of her photograph) and with her artificial vitalities (although the narrative culminates in signs of human reproductive success with the absurd "happy-family shot" of pregnant Ruby making brownies).

This unlikely heterosexual couple—with her hyperfemininity and his understated, hesitant masculinity—share the place of conductor across the human-nonhuman axis: as the musical director of the photocopy machines (in between botching copy jobs, Sandy conducts the rhythmic sounds of the photocopiers as if collectively they were an electronic orchestra); as the transmitter of live electrical currents, Ruby's gestures raise and lower the brightness of the lights in Sandy's bedroom. In this deadpan bedroom scene, when the electricity between them literally becomes apparent, Sandy conducts Ruby conducting the current in the room, as if it were music. Ruby's ability to vary the charge and alternate the current turns her into a transductor, defined in relation to electricity as "a reactor . . . having a d.c. winding to control the saturation of a core and an a.c. winding whose impedance is thereby changed, so that a small change in direct current produces a large change in alternating current" (OED, definition 2). Underscored by the electronic music on the soundtrack that accompanies the variations in current, her unconventionally affective body translates light into a tonal register. In this sense, *Teknolust* takes Baudrillard's nightmarish scenarios and transforms them into comic form. The transductive circuits of shifting technical communication between living and nonliving actors meet in the increasing inextricability of human and nonhuman reproduction and replication. Pulling in the opposite direction from Baudrillard's bleak predictions, *Teknolust* celebrates precisely what Baudrillard's account most fears: the loss of the singularity of the human, the redundancy of embodied sexual reproduction, and the threat of not being able to see difference in sameness.

The transductions and biomediations in *Teknolust* make tangible those forms of vital connectivity that usually remain elusive: the seductive power of Hollywood heroines, the electricity of a romance, the patterns of viral infection, and the transmission of genetic traits. Giving shape and definition to the insubstantial, the film performs a series of pleasing materializations, which both satisfies and mocks our desire for tangibility by showing the digital clones as absurdly hyperintelligible bodies. In dispersing vitality across these multiple processes that unite human and nonhuman actors and integrate digital and genetic forms of life, the film compensates for the loss of specificity and singularity by letting us sense the potentiality of new generative configurations.

Yet there remains a tension between the film's formalization of cloning, through which it transforms the genetic imaginary into a series of aesthetic puns, and the lingering sense of things that cannot be resolved through scientific or cinematic technique. The yearning for authenticity that surfaces intermittently throughout the film is only partly met by its witty technical conversions, leaving unfulfilled something that is more enigmatic; the possibility of an ethical sexual relation is posed and transposed but never made intelligible; the importance of affect, emotions, and feelings are connected to other more substantial vitalities but not contained by them; the intensity of the search for meaning and purpose reaches beyond traditional humanisms, but also sometimes returns us to them. The tidy color-coded clones and the ingenuity of their circuits of exchange and transformation offer a sense of order and control that never fully contains the challenge their existence poses.

There are a number of scenes in the film where a cliché seems allowed to linger in a nonparodic form for a moment, which both articulates these yearnings and leaves them unresolved. In the rooftop scene, Ruby and Sandy exchange meaning-of-life thoughts, looking out over the cityscape with the wind blowing in their hair. Leaving behind the deadpan humor of the rest of the film, they move into a more serious style of conversation and agree that life's true meaning lies in art's capacity to move us—in our response to a Man Ray photograph or a Louis Armstrong number. In the jazz club scene, the liveness of the black singer Paula West (playing herself) performing on stage provides the immediacy and presence that the digital clones search for but, as in the rooftop scene, suggests that these qualities cannot be easily captured and reproduced. We are left with the sense that authenticity may seem like an outdated aspiration, but it cannot be dispensed with entirely. Neither can it be easily programmed or literalized: it remains ephemeral, ungraspable, and thus still desirable.

It is not only new forms of life that are at stake here, but also new modes of being alive: the ontology of aliveness. In offering an imaginative reconfiguration of the genetically engineered body's vitalities, the film posits an affective relation firmly within the technical domain, while playing with our desire for continued authenticity beyond it. *Teknolust* explores the desire to literalize affect through technology (both cinematic and genomic) but leaves us with a sense of the impossibility of such a project. In imagining the body as the medium and biology as the resource in the techno-scientific appropriation of an affective register,[17] the film rehearses what Marie-Luise Angerer has called the current "desire for affect" (Angerer 2007).[18] If affect signals that which the subject cannot fully grasp (precisely that which is beyond literalization)—but which reminds us we are social and subject to the impressions of others and refers to that intangible layer of embodied sensitivity or vitality which prepares the ground for the more concrete phenomena of emotions or feelings, but is not reducible to them—then the importance of embodying affect, rather than merely enacting or engineering it, is central to imagining cloned futures. No longer the property of the embodied subject, no longer the register confirming we are human, no longer that which distinguishes us from artificiality (whether celluloid goddess, robot, cyborg, or clone), the affective domain, as the final frontier of technical replication, here renders the body immediate through geneticization, that most organic of biomediations which disavows its technical origins even as it performs them.

Both the cinema and cloning articulate the desire to make the human body intelligible through technical manipulation, while simultaneously performing its immediacy. In blending the two through the citations and reiterations of transduction, *Teknolust* displays the mutuality of digital and genetic mutability and the converging histories of desire that have defined their formation. The multiple conversions of vitality, from celluloid to biological and from electronic to genetic, promise an unbounded life force that can be mobilized for infinite generative expansion in accordance with these desires. But something nevertheless escapes the formal pleasures of translating material agility into stylish composition. What remains is the sense that, despite these innovations or perhaps because of them, intimate connection with others may still escape us and may be beyond our technical control.

[9] Enacting the Gene

The Animation of Science in *Genetic Admiration*

What would it mean to think about the relationship between the gene and the image as both more and less than analogous? When I suggested in chapter 7 that the prospect of genetically engineering the human body may produce what I called a sense of lost bio-aura similar to Benjamin's famous notion of the lost aura of artwork in the age of mechanical reproduction, my argument operated through analogy. If the artwork can be replicated, it loses its aura of distinctiveness and presentness; if the body can be genetically engineered, it loses its aura of organic singularity (its bio-aura). In each case, a sense of authenticity is endowed on past relations at the moment of their loss. As I suggested, bio-aura might be something we feel only when faced with the possibility of techno-scientific interference with genetic processes. The theoretical speculations outlined in chapter 7 work only if the changing reproducibility of biology and culture makes some kind of analogous sense to the reader.

The current chapter is both an extension of, and in critical dialogue with, the arguments about the analogies between the biological and the cultural set out so far in this book (see especially chapters 3 and 7). In this final chapter, I read the changing status of the relationship between the image and the body as constituted by genetic and reproductive technologies in each of these overlapping spheres through a discussion of a short collage animation film titled *Genetic Admiration* (Frances Leeming, 2005). Using techniques of animation that transform both the image and the body, this film moves beyond analogy toward enactment; or rather, it produces enactment through its analogies. In a burlesque combination of images of Walt Disney World, fairgrounds, expositions, advertisements, and Hollywood stars, as well as the visual and musical styles of several genres (the melodrama, the western, and the natural-history film), *Genetic Admiration* reconfigures twentieth-century entertainment cultures into a spectacle of artificiality and commodity display, extending the impulse to the microscopic activities of reproductive cell life: the gene. In what follows, I investigate how the film uses a combination of collage and animation techniques to make us question our sense of "genetic admiration" and how, in so doing, it explores the associative iconographies and narrative landscapes of the genetic imaginary.

Collage animation cinema combines two distinct image-making practices especially appropriate for a critical exploration of technical interference at the genetic level: collage depends upon collecting and assembling preexisting or used materials and animation gives life to inanimate objects or figures through assisted movement. Blending the recomposition of diverse fragments with the production of lifelike action or movement is immediately suggestive of the moves of reproductive interference through genetic engineering (such as the transfer of a nucleus from one egg cell to another, extracting cells from one body to use in another, or combining human and animal DNA) and of the life-generating potential of engineering human cells, organs, and bodies (in work with stem cells, embryo transplantation, and cloning). But this is not to suggest that the technical moves of this collage animation simply seek to mirror the recombinant art of genetic science; rather, *Genetic Admiration* combines these formal devices to draw attention to the problems with the promise of scientific miracles and to mock our current worship of heroic geneticists. Scientists now promise to imitate in a laboratory the regeneration of human life, and this film's animation techniques are used to show how such promises fit in a long history of spectacular innovation and global displays of pride in scientists as national heroes.

Collage is based on the idea of "making pictures by sticking together bits and pieces of random and miscellaneous bric-a-brac which might take one's fancy and stir the imagination to release hidden associations" (Wolfram 1975, 7). What better form through which to explore the genetic imaginary than an artistic practice premised on free association? The collage work in *Genetic Admiration* exposes just such a level of signification—those apparently irrational, hidden but traceable chains of connections and associations between things which surface in unexpected though repeated forms. The film combines this with an animation form of manual manipulation shot in real time, so that the collage assemblages perform the human labor of bringing things to life in a way that invites reflection upon the mechanics and agency of these life-giving interventions.

Each move in the collage animation techniques of *Genetic Admiration* is made manually in real time, turning the filmmaker into a puppeteer. The cuts that literally enable motion (the dropping of a character's jaw to simulate speech, for example) are made visible to form part of a mise-en-scène of recombinant assemblages. The unfamiliar slowness of watching real-time movement on the screen causes us to pause, while parts of the mismatched bodies move awkwardly within the frame. Like the geneticists in the laboratory, the animation artist seeks to breathe life into new forms of matter (the word "animation" comes from the Latin *anima*, meaning breath or soul). But in this film, life is not breathed invisibly into matter, nor is form magically given life; rather, the laborious work of manipulation is made as cinematically present as possible, in contrast to the seamless moves of digital animation films (which hide their techniques) and the invisible artifice of genetic engineering (whereby you literally cannot see the difference between Dolly and other sheep, or that between a test-tube baby and other babies). This literalization displays the formal work of image assemblage and pushes against the genetic aesthetic of films like *Code 46* (Michael Winterbottom, 2003), *Gattaca* (Andrew Niccol, 1997), and *Teknolust* (Lynn Hershman-Leeson, 2002), with their mise-en-scènes of smooth surfaces, repetitive sequences, and pleasing symmetries (see the earlier chapters on these films and the introduction).

In *Genetic Admiration*, we cannot ignore the animating work of making something lifelike, of moving it to action; neither can we overlook the human agency and desires behind such labor. Through its insistent and awkward presence on the screen, the manual work of animation becomes precisely what we observe. Through these performative techniques, the film reveals the ellipses that are so often concealed in the cultural production and circulation of a spectacular science like genomics. Thinking through the extent to which technologies of visualization and of embodiment might be both more and less than analogous, this chapter explores how this film performs the process of cinematic animation as an interrogation of the desires permeating genetic engineering and cloning.

Genetic Admiration might be best described as a critical bricolage that pillages and reassembles familiar sounds and images from mass culture. The film falls into three distinct, overlapping parts: the Disney phantasmagoria, the visit to the Centre for Genetic Admiration, and the turbulent trip to SeaWorld. Running through all three segments is an examination of the architectures of display which have shaped the mass circulation of scientific ideas about evolution, generation, heritage, and progress. Placing the contemporary celebration of genetic experimentation within a collaged mise-en-scène of theme parks, fairgrounds, Expos, and Hollywood star cultures, the film frames our response to the miracles of science through its deconstruction of forms of mass entertainment and highly recognizable popular pleasures. Exploring how scientific discoveries such as cloning and mapping the human genome are embedded within twentieth-century Western consumer cultures, these overlapping histories are enacted through the decontextualization and recombination of familiar figures, heroes, and icons. Sliding from one set of free associations to another (genetics—clairvoyance—séance, or Fordist production line—beauty contest—sperm/egg donation) these image sequences bring together unlikely collections whose irrational and unexpected connections foreground precisely the absurd logic that blends together capitalist cultures and scientific endeavors in the name of genetic admiration. Through its refusal of the smooth and swift techniques of digital animation, the film explores the genetic lifeforms of cloning cultures as it slows the speed of the suturing of genetic engineering into our everyday lives. The genetic imaginary that binds together these chains of association is undone.

To begin to undermine the force of the conventions through which popular science has naturalized and neutralized its own figurative field demands attention to the detail of cultural form. *Genetic Admiration* puts the conventions of popular culture to work, transforming their wisdom into comic absurdity and pushing the structuring desires of science to the surface through humorous excess. Techniques of collage animation place the current wonder at genetic engineering within the history of the public display of the monsters and marvels of science and nature.[1] Not sharing the genetic admiration which surrounds us today, the film explores how the imaginary of the science of cell biology, which intervenes in the microactivities of life itself, draws upon and extends existing masculine fantasies embedded within the cultures of popular science.[2]

Whenever anything in this collage of popular genres and tropes in *Genetic Admiration* appears to fit together neatly, the moment soon passes; the awkwardly animated joins quickly become the film's signature style. The disproportionate scale of figures and ground produce a disorienting and often comic effect: Walt Disney's large, grinning head on the tiny body of a bee makes his pollinating desire absurd (see figure 40); the movement of a roller-coaster car away from its tracks turns the accompanying screams of delight into signs of terror; a gigantic red molecular structure tilts back and forth at a fairground and performs with bizarre self-satisfaction. The emphasis on artifice throughout is reinforced by the ambiguous status of the diegetic components: are they still images or moving ones, and if they do move or are moved, what does the movement imply? While some waving arms and walking legs suggest automation, their disjointed attachments to the rest of the body or the mismatch of the body with the ground beneath its feet produce animation in the most jolting, awkward sense of the term.

What we might call an animation of disjuncture also appears in the contrast between the use of black-and-white and color photography in the film. The exaggerated tonal contrasts of the Technicolor Ontario Place fairground and Disney nature park return us to a previous era of cinematic brilliance, where early attempts at color now look unnaturally dazzling. Refusing to provide diegetic coherence or familiar temporal orientations, two styles of photography and filmmaking coexist within the frame of dialogic commentary: the black-and-white sections suggesting a time before color film, and the dazzling combinations of Technicolor invoking an outmoded 1950s cinematic style. Both refer in

40. Walt Disney as a pollinating bee in *Genetic Admiration* (Frances Leeming, 2005).

different ways to a bygone era. As abstract ideals are condensed into familiar scenarios, gestures, figures, or tropes, the Past as snapshot image is paraded before us, alongside Civilization, the Pioneer Spirit, Nature, Progress, and of course, in the promise of genetic science, the Future.

Manual manipulation is everywhere in evidence in the inappropriate scale, proportion, connection, coordination, and flow. The seemingly unskilled, childlike, stilted cut-and-paste collage techniques foreground their own mechanisms even as they invoke retro styles of stilted puppetry which once entertained us on television (like puppets of the 1960s in *Stingray* [Alan Pattillo et al., 1964–65] and *Thunderbirds* [Desmond Saunders et al., 1965]). Along with the absence of synchronicity, only selected components of the shot are animated: a single person in a crowd, one car on a roller coaster, or even just a leg, arm, or jaw of one of the figures. Together with an insistence upon using real time, the stillness surrounding the animated parts contributes to the sense of discordant motion. These elements combine to deny us the speedy pleasures of mainstream animation and familiar flows to which we attach a sense of agency and progress, and which characterize popular television programs about science. As our desire for a technological future which includes us in its fluid transitions and transcendent speed is thwarted, we are left with a collection of images whose controlled and selective animation requires us to reflect critically on these very desires.

With its deployment of these techniques for the purpose of political critique, *Genetic Admiration* is reminiscent of some of the early avant-garde collage and photomontage aesthetics of Hannah Höch and John Hartfield, among others. Just as Berlin Dadaists used collage and photo-montage to ridicule the pretensions of an outmoded political regime and its absurd leaders, Leeming deploys similar strategies to undo the conventional constructions of genetic science as an object of unques-tionable national and international pride, and of geneticists as heroes. Dadaist techniques of free association bring together the film's bizarre assemblages of sounds and images of technology and scenes of the ad-miring consumption of scientific achievement. The influence of sur-realists such as Höch (1889–1978) on Leeming can be traced through a similar visual style that depends upon deconstructionist techniques for political critique. One critic has suggested that Höch's style poked "exu-berant fun at the anachronisms of the decaying feudal order of Kaiser Wilhelm and the newly-emerging age of industrial juggernauts" (Wol-fram 1975, 82). Like Höch, Leeming draws upon a highly diverse archive and attaches humor through associative critique with her rescaled and recontextualized bodies. In Höch's "Cut With a Kitchen Knife Through the Last Weimar Beer-belly Cultural Epoch,"[3] for example, "the Kaiser, Hindenburg and the Crown Prince [are] dressed as chorus girls, while the new emerging leaders of Weimar, Ebert and Schlacht, with their con-sorts, flash through Valhallah-like space alongside acrobats and sports stars, all in the midst of exploding machine parts" (ibid.). Maud Levin suggests that "in the most general sense, all of Höch's Berlin Dadaist photomontages can be read as cuts through the contemporary scene, a filmic cataloguing and recombining of signs of modernism" (Levin 1993, 8). When juxtaposed, these fragments offer allegorical interpretations. In her careful readings of the different phases of Höch's work, Levin argues that montage is central in considering "feminine representation and its ambiguities," and in exploring how significant the female body was to the political and cultural changes of Weimar Germany (ibid., 9).

Unlikely combinations of images from politics, popular entertainment, and technology appear in *Genetic Admiration*, which is similarly preoccupied with how skewed bodies and buildings and incongruous suturing effects might produce a political critique. Like Höch's collage work, *Genetic Admiration* uses reassembled and proportionally mismatched bodies as a form of cultural and political satire. The film is especially critical of the current fascination with the gene as part of a broader investment in the microscopic truth of the human body and its genealogical traces. The film's strategies also follow Höch's style by introducing what Gunda Luyken calls the "disturbance factors" of avant-garde collage, the element which does fit or interrupts the themed flow across the collection (Luyken 2004, [5]). But in *Genetic Admiration* there is not a singular disturbance; rather, familiar reference points mutate in unexpected directions, making unlikely associations.

The film's collage form reiterates the centrality of collecting and reassembling to the visual pleasure involved in the public consumption of science, and also literally takes apart the desires that attribute spectacle status to such collections. Playing upon this paradox, the film displays an admirable archive of photographs, sounds, and images and questions the pleasure driving these assemblages. Collage itself is always a citational artistic practice, since it uses second-hand components; *Genetic Admiration* amplifies this effect through its imitation of earlier combinations of skewed and reassembled bodies in avant-garde collage. Since one of the film's repeated concerns is to locate current trends of popularized science within a history of their visual and spatial organization (however scrambled these recombinations might feel), *Genetic Admiration* presents us with a series of collage forms which reference their own history of citation. As a science of engineering cell components, extracting them from the material of one cell and injecting them into another, cloning might also be thought of as a citational practice, a kind of biological equivalent to social disembedding and reembedding.[4] But unlike the sutures shown in the film, those of genetic engineering and cloning typically remain invisible.

In the first segment of *Genetic Admiration*, we move from aerial photographs of Toronto's cityscape, showing the famous tourist sites of the CN Tower, City Hall, and Ontario Place[5] through cut-out stills of dated fairground gaiety to the comical image of Walt Disney's head grafted onto the tiny body of a busily pollinating bee. As a black-and-white photograph of a woman—Joan Crawford, as Mrs. Disney—offers a tearful narration of her sorrows to the figure of a fortuneteller or psychotherapist with a crystal ball and a couch—Marlene Dietrich, who travels round the fairground in a brightly colored Romany cara-van—(see figure 41), Walt the bee buzzes gleefully among the flora and fauna of his Technicolor natural world, leaving his generative mark on each flower that he visits. Like the heroines of 1940s melodrama who see only a bleak future for themselves—in a mode typical of what Lauren Berlant has called the culture of the "female complaint" (Berlant 2008)—this weeping woman laments that her husband "calls himself the pollinator but he can't even pollinate his own wife." Her despair contrasts with Walt's insouciant cheerfulness and profligate pollina-tions, through which he guarantees himself a place in the future by way of his reproductive generosity.

This hybrid figuration of Walt the bee's desire to be present in the fu-ture connects Disney's public role in producing utopian fantasies about an idyllic nature for young audiences with his private desires for science to conquer death (as is well known, he wanted his body cryogenically frozen). It is a short step from his personal desire for technological res-urrection to the fantasies of immortality that motivate the techniques of genetic engineering and cloning. Writing of the Walt Disney Com-pany, Baudrillard suggests that:

> The Disney enterprise goes further than the imagination. The great pre-cursor, the initiator of the virtual reality of the imagination, is currently taking over the whole of the real world to incorporate it into its synthetic world in the form of an immense "reality show," in which it is reality itself which presents itself as a spectacle, in which the real becomes a theme park . . . Disney should . . . buy up the human genome, which is currently being sequenced, to turn it into a genetic attraction. (Baudrillard 2002, 151–52)

As an iconic figure of the twentieth-century entertainment industry, Disney belongs to the realm of immortality not only because hundreds of films carry his signature for posterity, but also because of his own famous wish to avoid the finality of death. Baudrillard asks: "Why not cryogenize the whole planet, exactly as Walt Disney had himself cry-onized in liquid nitrogen, with a view to some kind of resurrection or

41. Marlene Dietrich as a fortuneteller or psycho-therapist with a client, Joan Crawford, playing Mrs. Disney in *Genetic Admiration* (Frances Leeming, 2005).

other in the real world?" (ibid.). Such a wish is futile, not because of the limits of science but because

> there no longer is a real world, and there won't be one—not even for Walt Disney: if he wakes up one day he'll get the shock of his life. In the meantime, from the depths of his liquid nitrogen he goes on annexing the world—both imaginary and real—subsuming it into the spectral universe of virtual reality in which we have all become extras. The difference is that, as we slip into our data suits or our sensors, or tap away at our keyboards, we are moving into living spectrality, whereas he, the brilliant precursor, has moved into the virtual reality of death. (Ibid., 152)

Baudrillard's apocalyptic pronouncements are caricatures of contemporary cloning cultures (see chapter 1), but his association of Disney with the aspirations of the Human Genome Project makes a number of connections similar to those explored in *Genetic Admiration*, especially the ones between the manufacture of desirable natures, Disney's role as a producer of animation, and the potential engineering of life in the genetics laboratory. These fantasies of generating life cut across the cultural forms uniting science and popular media, tying us into the heroic fantasies of masculine adventure narratives about crossing frontiers, conquering new terrain, and revealing the hidden truths of nature in all her mysterious guises.

The image of Walt Disney as the pollinator in *Genetic Admiration* brings to mind numerous other national heroes who embody a sense of excitement about their role in the technological possibilities of the future (see McNeil 1987). In Britain, one such celebrity is the famous in vitro fertilization specialist, Sir Robert Winston (now Lord Winston), the narrator of the much-publicized BBC science documentary series *The Human Body* (Christopher Spencer, 1998). In the second episode, "An Everyday Miracle," he peers through a microscope at a sperm sample. An image of wriggling sperm is then accompanied by the following voice-over: "These are my sperm. Amazingly, about 500 million of them from a single ejaculation. With just this one ejaculation it should be possible to impregnate all the fertile women of Western Europe. And I'm nothing special" (Winston 1998). As a doctor who has "fathered" hundreds of so-called test-tube babies (their photographs cover the bulletin board on his office wall, frequently shown behind him in interviews), Winston's statement about the reproductive potential of his own semen combines the authority of the neutral scientist and documentary narrator with the desires of a would-be surrogate father and the reputation of a highly successful practitioner of in vitro fertilization, revealing an overarching fantasy of himself as the universal male inseminator (at least for Western Europe). This display of masculine prowess in both scientific techniques (magnification is only the beginning in the series, which promises to take us on a journey inside the human body using the latest visualizing technologies) and the wonders of nature (imagine that many sperm in just one ejaculation) confirms the centrality of fantasies of sexual and reproductive potency to the supposedly natural facts of life which science as culture presents to its public. Popular science programs such as this (often narrated by celebrity scientists like Winston) promise to show and tell audiences about the latest inventions and discoveries in an accessible and supposedly neutral way, while rehearsing what many feminists have seen as a clichéd set of normative values about sexuality, gender, reproduction, the family, and the nation.[6]

If the Disney bee amuses us as the epitome of the masculine desire to be seen as the agent of fertility, it does so because we recognize its pervasive presence in contemporary popular representations. When the image of Winston in *The Human Body*, admiring his propagatory potential in the name of educating the public about the miracle of life, came into my mind while I watched the buzzing Disney bee, the association arose because of the shared masculine fantasy which underlies this proud display of fecundity. Through their delight in the technologies of reproductive display, Winston and Walt the bee enact the familiar and highly conventionalized masculine desire for generative agency and procreative glory. But it is not only the parodic ridicule of these familiar masculine desires which might make us think of conventional BBC science programs, it is also the critical dialogue with the visualizing techniques of this genre which *Genetic Admiration* instantiates through its formal experimentation. The cut-and-paste figures of this animation film speak back to the smooth, digitized flow of a television science program like *The Human Body*. Refusing the seamless transitions and mutations of digital imaging, which have become so familiar to contemporary viewers of television and cinema, *Genetic Admiration* challenges the sedimented forms of knowledge, the worship of technological innovation, and the new modes of corporeal perception such as media convergence achieves. Documentaries like *The Human Body* rely on combining the latest medical discoveries with new imaging techniques to offer the viewer a sense of a spectacular and adventurous journey inside the human body (in particular, inside the reproductive female body). Advertising itself in precisely these terms, the digital techniques used here allow a seamless movement from the body's surface to its innermost depths in seconds: shots of a naked female body morph into close-ups of her ovaries and even her egg production as the viewer travels inside a corporeal landscape which promises to reveal the secrets of the perilous journey of birth. The rapidity and flow of these transitions from corporeal exterior to interior and back again give the viewer the feeling of traveling into the body's hidden spaces and microscopic secrets.

The visualizing technologies of such popular science documentaries offer the pleasures of what Kim Sawchuk has called biotourism: "the persistent cultural fantasy that one can travel through the inner body, a bodyscape which is 'spatialized' and given definable geographic contours" (Sawchuk 2000, 10). As a result of the growth in medical techniques for "imaging the interior of body space," which make the "inside of the body into a landscape for imaginary travel," Sawchuk argues that "the biotourist" emerges as a new identity produced by "the medicalization of the body as an administrative unit" and the transformation of our "corporeal sensibility" (ibid., 20). Programs such as *The Human Body* capitalize on just such a sensibility: they take viewers on a voyage inside the human body with the very latest of imaging techniques. As

Sawchuk argues, such forms of figurative travel are dependent upon a newly imagined sense of interior bodily space (with plenty of room for a visitor to see what is happening), which has been produced by medical imaging techniques and distributed through popular forms of entertainment:

> Contemporary medical technologies such as microscopes, X-ray machines, endoscopes, and most recently positron emission tomography (PET) and magnetic resonance imaging (MRI) have made it possible for medical scientists to peer into the body without cutting through the skin . . . these devices *visualize and enlarge somatic space*, rendering our most infinitesimal cells, molecules and genetic structures into images on a scale that we can more easily comprehend. (Ibid., 9, emphasis added)

Offering a critique of precisely these pleasures of biotourism, *Genetic Admiration* explores the ways in which display and spectacle lie at the heart of both the glorification of science and the popular pleasure of mass entertainment. At this nexus, the desire to see and the desire to know have become inextricable, and the massive scale and circulation of the spectacle is matched only by the increasing belief that the essential truth of life lies in the miniscule, the minute, or the microscopic (see Roof 2007). Drawing on Susan Stewart's work (1993) on the "literary legacy of the miniature and the gigantic," which argues that "the secret of the microscope is its transformation of the miniature, which can be viewed with a single perspective, into the gigantic, which can only be taken apart piece by piece," Sawchuk argues that "this question of scale and the inversion of the microscopic and macroscopic is a central feature of the fantasy of biotourism" (Sawchuk 2000, 11–12). What Sawchuk calls "this transposition of scale" prepares us, she argues, for "the spatialization of the inner body and its transformation into landscape" (ibid., 13). *Genetic Admiration*'s "transposition of scale" works in conjunction with its critical deconstruction of the conventional pleasures of biotourism so eloquently described by Sawchuk. Throughout the film, disproportionate assemblages enlarge human bodies or body parts to absurd scale: a giant, chubby baby's foot appears on a number of occasions, including on a séance table surrounded by men in business attire (see figure 42); a giant, headless male body is displayed upside down in front of Toronto's City Hall (see figure 43); and as part of the fairground display in Ontario Place, inappropriately oversized bodies

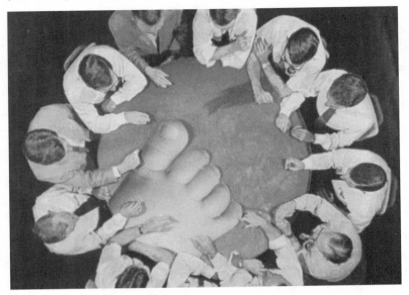

42. Men inspecting a giant baby's foot in *Genetic
Admiration* (Frances Leeming, 2005).

are on show next to advertisements for other freaks of nature and sexual
perversions in true circus tradition—"In Person, the Third Sex Family,
The Francos" and "Stranger than Rosemary's Baby." In the Disney garden idyll, Walt the large-headed bee follows a group of 1950s-style, perfect white American boys who plant and fertilize tulips that instantly
grow to gigantic proportions, eagerly awaiting pollination from the hybrid Disney insect. Fairground mysticism and fairy tale blend together
here to produce bizarre surrealist mutations and recombinations: a vinyl record of "It's a Small World"—the song played at the ride of the
same name at several Disney theme parks—mutates into a table top
around which (in a combination of business meeting and séance) sit
men in white shirts and ties, gazing down at a mysteriously huge baby's foot; a large egg rolls down a misty mountainside, is fertilized by a
gigantic male hand shaking a salt cellar, cracks, and opens to spawn a
tiny pioneer wagon labeled "DNA."

Not only do the miniature and the gigantic have a long literary and
cinematic legacy, as Stewart and Sawchuk document,[7] but they also
have a set of resonances in the genetic imaginary, where shrinkage and

43. A moving headless sculpture in front of Toronto's City Hall in *Genetic Admiration* (Frances Leeming, 2005).

expansion have a particular figurative force. Mrs. Disney complains to Marlene Dietrich, the psychotherapist-fortuneteller, that when her husband disappeared (while working in the garden shed) he left a "tiny note on his [discarded] clothes saying 'Please Just Let Me Bee,'" reminding us of one of the earliest films about genetic engineering, *The Fly* (Kurt Neumann, 1958), in which a "family man," driven by a passion for science, experiments with the molecular structures of nature and ends up half-human and half-fly, as a result of an accident in his laboratory. The hybrid Disney bee brings to mind the famous final shot in the earlier film, in which the miniaturized head of the scientist sits on the body of the fly caught in a spider's web, calling "Help me!" in his tiny voice as the spider approaches to devour its prey. The Disney bee references the long history in science-fiction cinema of fascination with the horrors of genetic interference and its potential hybrid cross-fertilizations and resulting distortions in human proportions.

Höch's description of her fascination with the collage form directly raises questions of scale: "I wish to demonstrate that little is also big and big is also little, and that the only thing that changes is the standpoint from which we judge so that all our human laws and principles lose their validity" (Höch quoted in Wolfram 1975, 82). Borrowing this logic of reversals, *Genetic Admiration* rescales our sense of proportion by using visual reassemblages to enact biological ones. Here the question of analogy enters the frame as the visible cut-and-paste animations articulate invisible biological ones. But the film pushes beyond a comparative mode through its performative aesthetic which enacts its own thematic, a procedure that J. L. Austin (1962) and Judith Butler (1990, 1993, and 1997) have described. Insofar as the film does not say what it does but does what it says, the relationship between the visual image and the gene is a kind of performative which plays on the cultural-biological analogy at stake but also moves beyond the paradigm of the parallel and into a filmic enactment of the genetic imaginary it interrogates.

Genetic Admiration explores our current sense of wonder at scientific innovation in the field of genetic engineering and cloning so as to place science firmly within the history of modes of capitalist production and consumption, and within forms of spectatorship of popular entertainment. Clichéd gendered pairings connect Hollywood genres and fairground fun: melodrama and fortunetelling; the western and rifle stalls. All three scenes of Marlene Dietrich as the fortuneteller-psychoanalyst satirize the conventions of Hollywood cinema. Dietrich is transformed from a 1930s Hollywood icon of feminine perfection and desirability (with her enigmatic aura and seductive, husky voice with its European accent) into a fairground fortuneteller whose clients have their psyches read and their futures predicted in her colorful Romany caravan, which makes the rounds of fairgrounds and exhibitions. Transposed to the era of genetic admiration, the mysterious Dietrich is animated to enact the female clairvoyant, who, like genetic predetermination, promises to see into the future for clients. As each client is invited to enter the magical world of fortunetelling, where you can sit and dream of a better life in the future, the curtains at the entrance of the caravan sweep back to reveal familiar characters from black-and-white cinema: the weeping woman, the suave entrepreneur, the damsel in distress. In the second consultation, Dietrich is animated to talk to her only male patient (a black-and-white cut-out of Gary Cooper), the bioentrepreneur, whom she reprimands in the clipped style reminiscent of classic Hollywood "battle of the sexes" dialogue: "you're arrogant, greedy, and not very bright, but I like my men like that." As they blow cigarette smoke across each other's faces (a gesture made famous by the closing scene with Bette Davis and Paul Henreid in *Now, Voyager* [Irving Rapper, 1942]), he claims to have the complete genetic code of Iceland in his pocket. "Four billion. Cold cash. Down. Yup. Blue-eyed blonde is mine," he boasts. "God help us all," Dietrich replies. "Whiteness as a miracle product in your pocket, you say." But "the future's not for sale," she warns him.

Plundering familiar genres and styles from twentieth-century popular culture, *Genetic Admiration* juxtaposes the universalizing norms such as the white, squeaky-clean, all-American 1950s family ideals mentioned above with the freaks and misfits who inhabit its symbolic borders (and its fairgrounds), like the gender deviants of the "third sex" family, the horrific promise of showing something "Stranger than Rosemary's Baby," or the fascinating multiple births of in vitro fertilization. A voice-over in the "roll up, roll up" style of the fairground announces, "Not only does little Jenny have a mummy, little Jenny IS her mummy," as we see multiple copies of a female figure in a business suit inviting the fairground crowd to come and look at the spectacle of cloning.

If the freaks of nature are related to the wonders of science, what better forum can there be for their joint display than the combined setting of the fairground and the exposition? As Stewart argues, whereas the "grotesque body of carnival" engages in a structure of "democratic reciprocity," the spectacle of the fairground involves a "distancing of the object and a corresponding 'aestheticization' of it" (Stewart 1993, 107). Stewart argues that the term "monster" has an etymological history related to both *moneo* (to warn) and *monstro* (to show forth), and contemporary monstrosity belongs to "the history of the aberrations of the physical body [which] cannot be separated from this structure of spectacle." Stewart continues with the suggestion that "the spectacle functions to avoid contamination: 'Stand back, Ladies and Gentlemen, what you are about to see will shock and amaze you'" (ibid., 108). Reproduction, especially the female reproductive body, has been pivotal in this spectacle of monstrosities. As I have discussed elsewhere, teratology refers to the study both of monsters and marvels, as well as describing the history of monstrous births in the field of embryology and of cancers of the egg or the sperm cell (teratomas) in oncology (Stacey 1997). The history of studying embryos to examine the basis of cell behavior, which has informed developments in genetic engineering and cloning, has placed the birthing capacities of the female body center stage as the object of scrutiny, fascination, and sometimes horrified amazement[8] (see chapter 2).

As nature's most popular miracle, birth—and its fascinatingly monstrous failed counterparts—has now been surpassed by the new miracles of genetic engineering and those of the visualizing technologies which popularize them for public consumption. In the second segment of *Genetic Admiration*, the exhibitions at the Centre for Genetic Admiration (where, we are told, "amazing things happen") invite the crowd to gasp at the latest "unnatural and unbelievable" wonders of scientific possibility: "a dead man can impregnate his living spouse; a grandmother can give birth to her own nephew; and a child can have five parents and still end up an orphan." The reconfiguration of reproductive kinship promised by genetic engineering is enacted by the mutational power of objects in the film, which the flow of the collage animates. Spatial transformations are enacted as architectures mutate into ridiculous scales and unlikely proportions. Miniature forms contain gigantic displays, and hordes of people are housed in microscopic spaces. A huge masculine hand cradles a cluster of tiny test tubes, labelled "sheep," "rabbit," "cattle," and finally "man"—in which the Centre for Genetic Admiration is housed (see figure 44). Dozens of tiny cut-out men in business suits ascend the staircase into the Centre (as if they had drunk the same potion as Alice in Wonderland) in search of a magical mystery tour, as genetic science is literally turned into mass entertainment: "have your donor cards ready . . . join us on a tour of the most advanced and least regulated reproductive facility in the world." Since science is now to be admired for performing magic, a display of genetic engineering and cloning is perfectly suited to this fairy-tale world of impossible dimensions—in which a tiny test tube contains an enormous exposition.

Throughout the film, the collage form enables one image to appear to almost give birth to another, as objects, icons, items, and figures emerge from within or behind one another (the huge tulips in Walt the bee's garden spawn children and Bambi-like deer; the giant egg produces the wagon). The magic of the cinema and of science coalesce and animation is shown to be their common business. Signs of this coalescence appear and disappear, evoking the representational techniques of both endeavors. In the background of the Centre for Genetic Admiration,

44. Tiny businessmen enter the Centre for Genetic Admiration through the test tube labeled "man" in *Genetic Admiration* (Frances Leeming, 2005).

levitating figures (reminiscent of those in Georges Méliès's early magical films) signify the convergence of the shared visualizing technologies belonging to the experiments with still and moving images conducted in the late nineteenth and early twentieth centuries.[9]

Architectures of display are juxtaposed throughout the film. Cutting from one image collection to another, the collage exaggerates our fascination with techniques of revelation: the curtains of the Romany caravan draw back slowly to show Marlene's parlor; and a portcullis is raised to reveal a painting of the Annunciation, to the soundtrack of "Stairway to Heaven." In opposition to the quick, smooth movements that make such techniques invisible through frequent repetition, the slow speed of real time and the jolting, deliberate moves draw attention to the mechanics of the process involved; the accompanying soundtrack of squeaking and grinding noises underscores this sense of the effort required to make these things move or to set them in motion. In some scenes, it feels as if all the technologies of the twentieth century are crammed into one frame: televisions, vinyl records, satellite dishes, airplanes, cars, and trains. The decontextualization of these signs of Western progress and their relocation within the discordant diegetic temporalities produce ridiculous combinations: a satellite dish on a traditional Romany caravan, or multiple cloned children astride horses on an old-fashioned carousel.

This carnival of visualizing techniques centers on our ongoing obsession with the spectacular. Modes of scientific scrutiny and spheres of vision line up to fulfill our desire to see life in more detail: the magnifying glass, the microscope, and the computer screen (perhaps the masculine scientific counterparts to the feminine mystical crystal ball). Inside the Centre for Genetic Admiration, a vast auditorium with a large stage at the front displays the achievements of different countries identified by numerous national flags, and a glass elevator full of people moves up and down the inside of the building, offering them a bird's-eye view of the spectacle in the exhibition hall. Resembling a gigantic lecture theater or cinema, this space becomes an Expo of Genomics—the organized display of the wonders of genomics. As the Dadaist free-association techniques intensify, absurd (yet strangely appropriate) connotations lead the viewer from the Fordist production line of Miracle Whip to the postmodern body commodification of egg and sperm donations or sales and DNA patenting. Test-tube species and triplet babies are displayed for public viewing, scientific measurement, and regulation, as giant eggs and sperm are blended together in industrial-size mixers fit for making cakes for the masses. The colonial and industrial legacies of desiring white perfection are revealed, as the shared ideals of family and nation are mixed and matched through familiar consumer styles, popular tastes, and formations of fantasy.

Writing of the history of expositions, Umberto Eco suggests that they have functioned for centuries as inventories in which "an enormous gathering of evidence from Stone to Space Age, an immense catalogue of things produced by man in all countries over the past ten thousand years, [are] displayed so that humanity will not forget them" (Eco 1987, 292). Eco draws on the work of Benjamin, who suggests that "world exhibitions are the sites of pilgrimages to the commodity fetish" (Benjamin 1978, 151). For Benjamin, these exhibitions are popular festivals which supply the foundations of the as-yet-unformed entertainment industry (ibid., 152). While their architects set out to "glorif[y] the exchange value of commodities," he suggests, they inadvertently "create a framework in which commodities' intrinsic value is eclipsed. They open up a phantasmagoria that people enter to be amused. The entertainment industry facilitates this by elevating people to the level of commodities" (ibid.). In enthroning the merchandise "with an aura of amusement surrounding it," Benjamin suggests, the melding of commodity fetishism and aestheticized display begins. As Eco puts it:

> A boat, a car, a TV set are not for sailing, riding, watching but are meant to be looked at for their own sake. They are not even meant to be bought, but just to be absorbed by the nerves, by the taut, excited senses . . . The merchandise becomes play, color, light, show. The objects are not desired in themselves, although the show is enjoyed as a whole; every wish is gone and what remains is pure amusement and excitement. (Eco 1987, 294)

But for Eco, up until the nineteenth century, expositions pointed to the past; it is only later that "the marvels of the year 2000 began to be announced. And it is only with Disneyland and Disney World that concern with Space Age is combined with nostalgia for a fairytale past" (ibid., 293). It is precisely this double move that *Genetic Admiration* enacts: Walt the bee inhabits a nostalgic nature of the past even as his genes take their place in the future; members of the white American family peer at their Stone Age equivalents on display behind a glass window, and a bond between the male in each family is formed across the centuries as they light the fire that symbolizes, respectively, survival and the pioneer spirit; the tiny pioneer wagon crosses the misty blue fairy-tale mountains, its DNA label scrambling linear temporalities; Marlene Dietrich is the fortuneteller from the past who visits the future in her crystal ball, producing intuitive visions instead of scientific facts. In her reading of EPCOT Center (named after the Experimental Prototype Community of Tomorrow, and now called Epcot theme park) at Disney World, Maureen McNeil writes:

> When it opened in October 1982, EPCOT had two components—*Future World* (the Disney exposition of technological progress realised through corporate America) and *Wild Showcase* (a Disney gesture to multiculturalism and global diversity). Disney World, the most popular tourist

site in the world, the paean to American middle-class white culture and
values thus *incorporates* (literally, given the corporate sponsorship and
perspective which sustains it) an exposition of technoscientific progress.
(McNeil 2000, 223)

For McNeil, the combination of "techno-triumphalism and techno-
tourism" on display here produces highly gendered forms of admiration
for national heroes as we are invited "to gaze amazed at American technological achievement and to marvel at the prospect of a technological
adventure" (ibid., 222). In a similar vein, *Genetic Admiration* places the
feelings described in its title within a history of entertainment forms
in which scientific practice has been commodified and turned into a
spectacle engendering a sense of national pride and of global awe and
wonder.

If, as Benjamin suggests, "the world exhibitions build up a universe
of commodities," then the human gene enters this space as the latest
in a long line of fetishized objects, now extended to the minutest par-
ticles of the human body which can be patented and are also for sale
(Benjamin 1978, 153). Cloning is placed within the historical iconographies of mass commodity production in the scene in which young girls
in 1950s-style party dresses try to stack numerous jars of Miracle Whip
into a huge North American refrigerator set against a background image of magnified sperm, and each child floats around like an astronaut
in space, lacking the grounding force of gravity (see figure 45). In the
Centre for Genetic Admiration, we are invited to think of parts of the
body as exchangeable commodities whose market value is increasing;
magnified images of eggs and sperm help us envisage our potential
consumption of each other's body parts. The voice-over descriptions
of the donors who wish to sell their reproductive services make them
sound like entrants in a beauty contest, finishing their accounts of "vital
statistics" (their nationality, race, height, weight, appearance, likes, and
dislikes) with lame refrains reminiscent of beauty-pageant contestants
who always say they want to travel, help people (especially children),
and get to know themselves better. But here the donors shamelessly announce their commercial motives: "I am a 34-year-old medical student
of mixed African-American/Dutch origin, 5 foot 9, 195 pounds, medium brown hair, blood group O positive . . . I want to be a sperm donor to help people have a family; the money I receive will augment my

45. A fridge overflowing with jars of Miracle Whip
in *Genetic Admiration* (Frances Leeming, 2005).

salary." The Centre for Genetic Admiration thus combines the techno-triumphalism (McNeil 2000) of the Expo with the commodification of the body at the international beauty pageant. In this flow of images connecting the endless identical jars of Miracle Whip on the Fordist assembly line with the sale of eggs and sperm on the open market, *Genetic Admiration* places reproductive genetic engineering within the spectacular history of commodity fetishism and contested notions of value.

The figure of the child as both guiding principle through the apparent chaos of mutating sounds and images and as himself or herself the site of transformational possibility recurs throughout the film; it is especially significant as a kind of perverse narrative agent in the Centre for Genetic Admiration. In *Figurations: Child, Bodies, Worlds*, Claudia Castañeda (2003) suggests that the figure of the child is characterized by its mutable potentiality, which articulates competing adult desires and projections. She finds something distinctive in how the category of the child operates as a cultural figuration: "While all categories, including that of the adult, can be deconstructed to expose the instability of their contours or borders, what is specific to the category of the child is the identification between the child and mutability itself" (2003, 2). According to this argument, it is not simply that "what is distinctive about the child is that it has the capacity for transformation" but also that "such a transformation is a requirement . . . the child is . . . never complete in itself" (ibid., 2). Consequently, she suggests, it is precisely "this incompleteness and its accompanying instability that makes the child so apparently available: it is not yet fully formed, and so open to re-formation. The child is not only in the making but is also malleable—and so can be made" (ibid., 2–3). Such malleability takes on particular significance in the case of genetic engineering, for it is a science of interference in cell biology and in the making of life in the miniature. *Genetic Admiration* reverses such projections, allowing child figures to play with their own agency and conferring disruptive potential on their activities. Gigantic enlargements (such as the baby's foot) and oversized children populate the film, placing the mutable child body ironically in dialogue with a wider scene of constant incongruous and unexpected transformations.

If the child for Castañeda is the object of adult projections, for Stewart the child represents the quintessential miniature. In contemporary culture, Stewart suggests, the child is a miniature of the adult who "continually enters here as a metaphor, perhaps not simply because the child is in some physical sense a miniature of an adult, but also because the world of childhood, limited in its physical scope yet fantastic in its content, presents in some ways a miniature and fictive chapter in each life history" (Stewart 1993, 44). Variations upon this notion of the child as a fantasy miniature adult resurface throughout *Genetic Admiration*: the Disney kids who romp innocently in "Natureland" alongside the enthusiastically procreative Walt the bee; the Miracle Whip girls whose premature domesticity goes out of control and destroys the public order and visual display of the Expo; the boys in the audience at SeaWorld in the pregnant woman's stomach, who pretend to understand more than they really do and end up in turmoil. To the extent that Stewart is right that "we imagine childhood as if it were at the other end of a tunnel—distanced, diminutive, and clearly framed" (ibid.), *Genetic Admiration* might be read as scrambling such clarity of vision and teleology of proportion, by repeatedly confusing the narrative linearity of both official histories and of personal life-course fictions, presenting instead

an image of turmoil, and cutting across the desire for meaningful genealogies and legible historical or cinematic perspectives. For example,
when the members of the clean-cut 1950s all-American white family
peer eagerly at the display of the Stone Age nuclear family, looking for
confirmation of their heritage and continuous genealogy, the father's
attempt to ignite a fire in collaboration with his prehistoric counterpart
transforms them all into a spectacle for another family from a later period (perhaps the 1970s), who look on curiously. As Walt the bee buzzes
across the frames of this sequence, gesturing toward fantasies of temporal and spatial continuity, the whole scene is commented upon by a
woman from yet another context: "that was a messy linkage between
procreation and heritage," she says to her boyfriend as she drinks the
bee, who has fallen into her paper cup of Pepsi (a quick and neat narrative solution to the problem of the overenthusiastic Disney pollinator
who doesn't want to die). Through these reiterative recontextualizations and reframings, the film thwarts both the desires for individual
teleology to somehow be a condensation of historical development (so
that biology can mirror history) and, relatedly, for the child to figure as
a miniature adult-in-the-making.

In a world where even DNA is for sale, commodity fetishism takes on
an accelerated and increasingly corporeal dimension. In the celebrity
cultures of both science and entertainment, feminine perfection bodies forth the masculine expertise whose ingenuity represents progress,
modernity, and the future. The desire to uncover the mysteries of the
female body has been seen to drive the unconscious desires behind both
modern science and the cinema. Whether as the smooth surface of the
screen goddess or the transparency of the female patient's physical interior, the body of the woman has been the object of knowledge and
pleasure, reflecting back upon the male egos of our national heroes and
their spectacular achievements.[10] The third segment of *Genetic Admiration* blends these two highly gendered, fetishistic traditions of visual
representation by transforming the domed belly of a gigantic cut-out
pregnant woman into a showcase pavilion, in which the wonders of nature are displayed to the crowds: the delights of SeaWorld, underwater
evolutionary excavations, and races between the yachts *Sea Section* and
Sperm Countess. As the woman lies obligingly still on her back (as if
in a traditional delivery room), with her vast, domed belly exposed to

the world like a background landscape for the city, her accommodating body encases the mysteries of nature, their deepest profundities passively awaiting discovery. The comic reversals of the miniature and the gigantic now become ways of exploring our disturbed sense of bodily integrity in the age of genomics. The fantasy of the reproductive body as a collection of improvable or replaceable parts is absurdly enacted in the multiplicity of sound and image components in any one frame. The diversity of origins and styles performs the cut-and-paste of the genetically engineered body and its transferable reproductive capabilities. The insistence on only a partial animation within the frame, while large sections remain still, enacts a sense of the selectivity and transferability of the making of life itself on the screen and in the laboratory.

Just as genetic material can be extracted and relocated, so one aspect of a sound or an image can be isolated and reused, recycled, or resignified through digital manipulation. Refusing the magic of both those techniques, *Genetic Admiration* asks: to whom do we attribute agency in such gendered fantasies of ubiquity, mastery, and omnipotence in this miniature world of microscopic wonders? This question of agency and structure is central to the film's exploration of the miniature. If the "miniature book always calls attention to the book as total object" (Stewart 1993, 44), then the film satirizes the idea—analyzed by, among others, Evelyn Fox Keller (1995 and 2000)—that the gene is a microcosm of the secrets of life. For Stewart,

> our transcendent viewpoint makes us perceive the miniature as object and this has a double effect. First, the object in its perfect stasis nevertheless suggests use, implementation, and contextualisation. And second, the representative quality of the miniature . . . becomes a stage on which we project, by means of association and intertextuality, a deliberately framed series of *actions*. (Stewart 1993, 54, emphasis in the original)

In this sense, the gene belongs to what Stewart calls "the daydream of the microscope: the daydream of the life inside life" (ibid.). As she explains, the miniature presents a "constant daydream," one in which "the world of things can open itself to reveal a secret life—indeed, to reveal a set of actions and hence a narrativity and history outside a given field of perception" (ibid.).

As many feminists have argued, the gene has become the latest iconic miniature in this long history of cultural fantasies of masculine mastery of human reproduction. Drawing on Keller's work, Judith Roof argues persuasively that scientific accounts of DNA "located the key to life's mysteries in the locus of the very small and structural" (Roof 2007, 3). In her own study of how ideas about the gene belong to a legacy of scientific investigation that is "heir to Enlightenment logics and nineteenth-century dialecticism," Roof suggests that DNA has become the "premier example of structure equaling function" and that we imagine "the DNA gene sharing in the same dialectical structures that we use to understand everything from language to kinship to law to history" (ibid., 32). Through linguistic analogy the gene is textualized and variously represents "the code, the book, the alphabet" (ibid., 7), and genetic engineering is reduced to forms of "textual manipulation (copying, correcting and amending)" (ibid., 32). According to Roof, "our idea of DNA centers the human as the agent of knowledge and the discoverer and decoder of a code that unfolds an orderly structure" (ibid., 32) and yet it also often ascribes agency to genes, "conceived as little personified agents or homunculi with wills and motives of their own" (ibid., 74). As Roof implies, with this attribution of agency, genes are less subject to the rules of biology than they are to the logic of narrative. Or we might say that they become subject to both in the conflated logic of narrative and biology.

It is this move which *Genetic Admiration* undoes so profoundly. If Roof is right, then the scrambling of scale and the disassembling of familiar modes of perspective and narration in the film might be understood to be in critical dialogue with such fantasies both of the gene as the most miniscule condensation of the meaning of life and of DNA as the reiteration of a meaningful structure, somehow expressive of narrative form. In *Genetic Admiration*, the spectator is denied the orienting sense of scale or narrative progression as the film pokes fun at both heroic fantasies of scientific exploration and cinematic conventions of narrative agency (whereby a singular, typically masculine, subject's desire moves events forward toward resolution), producing instead the chaotic sense of a shifting, transmuting collection of random changes. The desire to see inside the body, to take an imaginary journey through its interior contours, or to peer through a microscope to ascertain the truth of human history in minute biological structures are all reconfigured in these absurd manipulations of recycled images. Throughout the film, the sense of narrative agency shifts randomly, confounding our desire for linearity, meaning, and cohesion and attributing generative powers to unexpected and dispersed sources: who controls the sound of "It's a Small World," or the form of its mutating vinyl source? The relationship between the animating potential of cinematic and genetic techniques replayed here push the notion of a traveling nonhuman agency to ludicrous limits. If we search for answers to the question of who is controlling the next diegetic move, we find no reassuring distinction between human and nonhuman, or between living and inorganic. It seems that everything can now become an agent or an actor of its own animation in the world of autopoiesis (self-generation). But this fantasy is played against and alongside the laborious manipulations of parts or bits within the frame in such a way that the human participation in the animation of these images is ever present, though never directly visible. If the multitude of technologies and techniques appearing in the film poses questions about who can give motion and life to what and how, then the answer offered lies in another question: how should we understand the desire to animate in both science and the cinema?

In the final section of the film, underwater exploration is combined with aquatic spectacle, as the desire to conquer nature and reveal her mysteries is explored through a combination of familiar signifiers and tropes. A deep, confident, British sounding male voice delivers a David Attenborough-like commentary while a search for traces of ancient civilization through the portholes of a wrecked ship and along the seabed is conducted with utmost seriousness. Here the colonial project of Western expansionism, including its pioneering counterparts, and the growing reach of scientific inquiry into the depths of the sea appear to have some common drives: the sense of adventure in conquering new lands and taming other cultures and natures is reiterated in the revelatory impulses which take "man" deeper into nature (and the female body) in search of the truth of the universe. Just as the excitement of conquest motivated the ever-broadening reach of empire, so now science is motivated by the promise of knowledge to be found inside the body in that most condensed of forms, the gene. The current preoccupation with what we might follow Roof (2007) and call the truths minuscule, of which genetic science is only one dimension, places the history of science within its colonial context and then relocates it within the popular cultures of natural history displayed in the belly of a gigantic pregnant female body.[11]

If there is a single guiding figure at all in *Genetic Admiration* (and in many ways there is not), it is the androgynous Marlene Dietrich, who is the only character to appear in all three parts of the film, and who leads us across its highly gendered landscapes in her jolting caravan. In this third segment, the client who seeks her clairvoyant advice is the pregnant woman. Like a "desperate housewife," she confides to Dietrich that she "feels like a shipwreck": "I'm everybody's main attraction." When the Dietrich figure tells her, "it's time you rose up," the giant woman obeys. Still housing SeaWorld in her belly, she closes the lid to her interior and does indeed rise up and walk out of the frame, leaving behind her function as corporeal hostess to the spectacle of science and the wonders of nature. As she rejects the promised pleasures of Toronto's tourist sites and marches across the city, towering above its skyscrapers like a female equivalent of King Kong or Gulliver, she is accompanied only by the melodramatic orchestral music whose crescendos signal the imminence of a Hollywood ending (see figure 46). Does her move for autonomy and agency signal hope for the future? Is this woman, marching alone and determinedly out of the city, an optimistic contrast to the pervasive sense of commercial and conservative values saturating the normative cultures of science and mass entertainment in the rest of the film? Perhaps she knows that, in the age of genetic admiration, the future indeed is for sale, and she wants no part of it.

But with SeaWorld and its audience still inside her, there may be no escape for her. Her body continues to house the fantasy worlds of science and entertainment, and thus it may be that the materialization of normative desires cannot just be left behind by one rebellious figure— whatever the scale of her revolt. Perhaps there really is nowhere to go. However futile, the woman's bid for freedom at least turns the world upside down for a moment. It throws the bizarre contents of her belly into a chaotic mixture of spectacles and spectators who, like us, are jolted out of easy watching and denied the visual pleasures of smooth transitions and comfortable consumption.

Playing with the conventions not only of animation but also of cinematic form more generally, the film's overall narrative structure scrambles familiar modes of closure: in the first segment, the persistent demand for attention to the generative displays of masculine prowess end abruptly when Walt the bee is ingested in a large gulp of Pepsi; in the second, the genomics Expo is brought to a crashing halt when two conveyor belt lines of Miracle Whip jars traveling in opposite

46. The gigantic pregnant woman leaves Toronto in the
last part of *Genetic Admiration* (Frances Leeming, 2005).

directions collide in a disaster like a train crash; and in the third, the
freakishly gigantic woman rises up and leaves the city, creating turmoil
inside her. In an absurd reiteration of the formulas governing popular culture, each segment presents a narrative problem and a solution.
How to respond to the egos of male national heroes? Ingest them. What
could stop the commodification of even your genes? Proliferation and
collision. Who will save the woman's body from being the object of
masculine adventure narratives? She can cause plenty of chaos by just
getting up and leaving.

United by the scenes in Marlene Dietrich's caravan, the three segments of the film thus enact the repetitive structures they investigate.
As well as parodying conventional forms of narrative closure, the film
develops its own visual motifs, like the repetitive iconographies of particular Hollywood genres. These repetitions build a filmic mimicry
into the form that is already highly citational. In other words, there is a
double mimicry here in the recycled collage of familiar figures, icons,

gestures, designs, architectures, music, voice-overs, and settings (the weeping woman from the melodrama, the wagon from the western, or the roaming underwater camera from the natural-history film).

While the analogous moves are integral to the film's diegesis in the decontextualizations and recontextualizations of biological and cultural matter, these borrowings of second-hand sources from familiar genres of popular science and entertainment move beyond analogy into enactment, as an additional layer of mimicry becomes the film's structural signature. The film uses repetition to enact this double mimicry, citing its own representational gestures as a form of self-replication: the bird's-eye long shots which turn human bodies into repeating patterns; the moments of display and revelation, as doors open and curtains sweep back to transform everyday things into extraordinary spectacles; the spheres of vision (the crystal ball, the magnifying glass, and the ship's porthole); the promised knowledge of the future. Yet even as these formal repetitions remind us of the melodrama, the woman's weepie, or the underwater scientific adventure, guided by David Attenborough or Jacques Cousteau, these associations become fleeting moments in the endless transformation of form enacted by the film. This combination of citation and mutation generates a tension between recognition and misrecognition, as the spectator struggles to place the references (is that really the young Joan Crawford?) even as they mutate into something less recognizable. The citational strategy here, as in *Teknolust* (discussed in the previous chapter), is less a test of the accuracy of our memories and more a mimetic aesthetic that speaks to the genetic interferences it critiques.

Genetic Admiration undoes and remakes combinations of sounds and images that speak directly to the imagined practices of the science it interrogates: body parts are removed and supplemented; images are trimmed and expanded; scientific knowledge is reworked and relocated; and natures are reinvented and repackaged. Genetic engineering has been described as a recombinant science: it enables scientists to blend matter across species (human and nonhuman) and between the organic and the inorganic; to clone a sheep by transplanting the nucleus of an egg cell from one animal into a different host body; and to transfer embryos made from the eggs of one woman to another woman's body.[12] The perfect visual language for the exploration of a science which evokes awe and wonder at its fancy cellular footwork, collage animation both reassembles and brings to life: it disturbs the conventional filmic aesthetic of spatial and temporal coherence, redistributing body parts, voices, agencies, actors, and relationships. There is style and humor reminiscent of Monty Python in the instability of its multiple, simultaneous transformations and in the apparent randomness of their free-associative consequences. *Genetic Admiration* presents geneticized bodies through the carnival and the grotesque, in which the boundaries between human and animal, life and death, inside and outside are turned topsy-turvy in a restless world of reassembled forms. Unlike the smooth effects of digital animation and digital collage, which disguise their recombinant moves,[13] this piece draws our attention to the work of both techniques, making bizarre edits, ridiculous figures, and incongruous actions part of its self-conscious deconstructive aesthetic. We are left with a dizzying sense of the random (yet highly conventionalized) and arbitrary (yet clearly profit motivated) chaos of the current cultures of biotourism, and the film points out the absurdities of contemporary exchange values with which we are all becoming increasingly familiar, and which are rapidly becoming unremarkable.

To think of the relationship of body to image in the age of genomics as both more and less than analogous is to place the two reproductive cultures (the biological and the cinematic) in formal dialogue in such a way as to capture the echoes, but not to reduce one to the other. *Genetic Admiration* performs the moves through which genetic engineering promises to breathe life into new bodies, pushing beyond analogy toward enactment. We are left wondering both about the lack of fit between the two (the awkward joins of the film's collage) and about the space beyond analogy where formal performatives do their work at a different and more disturbing level. But the extent to which the film disassembles, even as it performs, its thematics may also be an indication of the ways in which *Genetic Admiration* is a kind of performative deenactment, a critical refusal of the pleasurable sutures which cover the desires of artifice. Without collapsing the one into the other, or merely displaying their analogous modes of aestheticized commodification, *Genetic Admiration* offers a visual enactment of the transformations of the relations of the part to the whole within both the cinema and science.[14]

Afterword

Double Take, Déjà Vu

The double take is an action prompted by the sight of something surprising or unexpected. It is an involuntary reaction generated by an arresting situation that passed unnoticed at first glance. The impulse to look for a second time hits in the spilt second of the turning away, as a sense of bewilderment transforms one movement into its opposite. The momentary delay holds two contradictory perceptions in tension: what I have just seen is unremarkable; what I have just seen demands a second look. Taken aback, I am literally taken back. A backtrack prompted by a perceptual disturbance, the double take feels like the mind catching up with the body, which has already reacted. The head has turned to see again, as the sense of the significance of the unforeseen belatedly enters our consciousness. The double take is both physical and psychic, conscious and unconscious, visceral and imaginative. What takes the subject unawares is an uncanny mismatch or the sense of "matter out of place":[1] the unexpected transformation of the appearance of a familiar person or object; a disturbing resemblance between faces; an inadvertent glimpse of a clandestine liaison; the sighting of something out of context; the return of someone from the past or even from the dead. This incongruent presence trips us up, interrupting the conventional flow of our perceptual coordinates. As uncertainty overtakes the mistaken first impression of normality, the subject is compelled to respond to the disorienting sensation by seeking visual confirmation.

The double take captures something of the character of the compelling disturbances that have been the focus of this book. But as an individualized response to a particular sighting, the concept is too narrow and concise to capture a more generalized reaction to the foundational interferences that have been of concern here. We might instead use the phrase "cultural double take," to refer to a shared sense of perplexity that generates the desire to look again at that which cannot be assimilated within existing perceptual habits and frames of reference. This is not a uniform response, of course, but a striking rehearsal of disturbance at particularly unimaginable prospects. The cultural double take refers to that which requires a second look within representational, as well as individualized, psychic formations.

The Cinematic Life of the Gene has explored the visceral force of the intangible disturbances occasioned by the prospect of techno-scientific interventions into biogenetic processes. The book has worked within the conceptual spaces between science and the cinema to identify the forms of their imaginary convergence. "Cinematic life" refers to the particular animating illusions of the medium that allow us to invest its fictions with a sense of vitality. The gene has life only insofar as it connects to wider dynamic circuits in the body. To give the gene cinematic life is to play upon a double illusion, and thus the title of this book has taken an impossible promise and turned it into a field of inquiry. Grappling with something ungraspable that registers through symbolic association, this book has tracked these connections and their broader cultural implications through the framing device of the genetic imaginary. The preceding chapters have elaborated the multiple ways in which such an imaginary is constituted through converging disturbances. Reading these disturbances across selected cinematic and theoretical texts, this book has investigated what we might call the cultural double take of the genetic imaginary: the sense of being taken aback by the prospect of ourselves that is somehow beyond us. A fantasy landscape generated by images, icons, scenarios, and discoveries, the genetic imaginary cuts across science, the cinema, and popular culture. Particular narrative and visual formations recur and mutate, borrowing and extending familiar tropes and motifs and inventing new associations and transpositions. Science fiction struggles to keep up with the latest laboratory inventions, and vice versa. These monsters and marvels hover on our horizons of possibility as we contemplate our own duplication.[2] The double take signals the ambivalence at stake here: our endless fascination with what might undo us.

The notion of the double take combines two concepts indicative of broader concerns at the heart of this project. In the preceding chapters, *the double* has appeared variously as the clone, cyborg, look-alike, impersonator, impostor, replicant, fraud, photograph, perfect match, adult offspring, identical twin, monster, and copycat. *The take* is an uninterrupted recording of something, such as a film sequence, that is almost always repeated: take two, take three, take four—until a satisfactory version has been achieved. Both an imperative and a noun, it registers the acquisitive desire of image making. To take a photograph is to apprehend something. Each film take requires, and in turn allows, a repeat performance. The double and the take are copies of different kinds; put together as the double take, they signal the necessity of repetition generated by something unnerving. Extended as the cultural double take, they combine a generalized sense of the interruption that prompts an involuntary return. The compulsion to repeat what we cannot absorb has us looking back for impossible certainties. In the face of our doubles, where are we to find ourselves?

The cinema has always presented its audiences with such dilemmas, insofar as its dreamlike animations take us out of ourselves and into the imaginary worlds of others. As Laura Mulvey puts it in *Death 24x a Second: Stillness and the Moving Image*, with "the arrival of the celluloid moving pictures . . . the reality of photography fused with mechanical movement, hitherto restricted to animated pictures, to reproduce the illusion of life that is essential to the cinema" (Mulvey 2006, 52). Sitting in a darkened room, surrounded by strangers (and perhaps with a few friends), watching the flickering images of a film projected at 24 (or thereabout) frames a second, feeling (if not hearing) it "prrrrrr" as it is projected onto the screen—these are the elements which combine to produce the ontology of classic spectatorship that we habitually incorporated in the celluloid age (ibid., 68). It is this particularly intense encounter with the moving image that has given rise to long-debated claims about the unique power of the cinema to stir our unconscious fantasies and to approximate an almost dreamlike state, inviting intro-jections and projections that have the potential to sometimes loosen (but also sometimes to confirm) our sense of who we are and what we want.

The cinema, Mulvey suggests, "combines, perhaps more perfectly than any other medium, two human fascinations: one with the bound-ary between life and death and the other with the mechanical animation of the inanimate, particularly the human figure" (ibid., 11). These fasci-nations prefigure what more recently have made the idea of techno-scientific interference into genetic activity so compelling: the extension of life beyond death through the technical animation of the human. There is no other medium that so closely anticipates the symbolic sig-nificance of the imitation of life that genetic engineering and cloning now represent. Cloning promises not just to confront us with an ex-ternal image of ourselves, or of another as duplicate, but to external-ize the phantasmatic self-other figurations which structure our inner lives and which animate the fears and desires upon which the cinema has worked its magic. From a psychoanalytic point of view, the dou-bling upon which the visual and narrative pleasures of the cinema de-pend attaches to existing relationalities of self and other (and of each in the other) that constitute inner psychic structures. The ego ideals on the screen perform us in seductive ways that we might invest in or we might repudiate. Either way, the form demands an attachment of the self as other and to the other (and therefore to the self) as image. We may later make it our own, remember it differently, or even use it to refashion ourselves; but whatever our contextual practices, the cinema continues as a system of duplication that depends upon multiple modes of doubling.

In its formal repetitions and obsessive returns, the cinema is the me-dium which turns psychic doubling into a visualization that demands a continual looking back: not the double take of the literal reflex, but rather the uncanny sense of presence and absence that typifies the cine-matic experience in its classical forms. If the structures of identification in the cinema work according to the disavowing logic of "I know very well, but all the same" (ibid., 11–12), they generate a strange sense of be-ing in between self and other that involves a kind of psychic doubling.

The uncanny of the cinematic image is conjured by these absence presences, whose status invokes an encounter with uncertainty. As Mulvey observes, for Freud the uncanny is always the return of the repressed old into the new, the recognition of the all-too-familiar in the unfamiliar, and it is the abject maternal body that represents the "truly ancient for the human psyche" and is the "true site of the uncanny." As she puts it, "not only is it the 'first home' itself but its once-upon-a-time memory of security and totality must become abject for the child to become an independent and autonomous being" (ibid., 50). As the former *Heim* (home) of the unheimlich (the uncanny), the woman's body represents an origin as unthinkable as the finality of death, which the subject struggles to comprehend fully. The uncanny of the cinema's animating potential that disturbs the boundaries between life and death, presence and absence, and organic and technological thus returns the spectator to the troubled borders of self and other that haunt the subject's sense of autonomy. As feminist film theory has elaborated, when the cinema animated the female body on the screen, it did not simply produce a reassuring spectacle of surface perfection—it also displayed its centrality to the phantasmatic ground of this mode of representation. In the cinema's animating illusions, it is not only the blurred distinction between the organic and the inorganic that disturbs and fascinates us, it is the duplicity or the "uncertain nature of femininity itself" (ibid., 51).

In the uniquely cinematic encounter with the screen image, the boundary between self and other, here and there, or now and then is a permeable one: there is a sense of both being and not being that person on the screen; of both occupying the time of the diegesis and that of the audience outside it. In the interplay of cinematic desires and identifications, tropes of loss and retrieval depend upon an imagined extension of subjectivity: we find ourselves in others; we lose ourselves in their fictions; and when we return to ourselves, we are transformed and yet still the same. Those longed-for moments of filmic closure both satisfy us through resolutions (so often unattainable in life) and leave a lingering melancholia for another time and place, and perhaps a yearning for another life that might have been, that we might have lived in a newly imagined elsewhere.

What Kaja Silverman has called the "excorporation" of cinematic identification foreshadows what techno-science has literalized in its practices of genetic transposition. Precisely because of the "real and the metaphoric radiance of the cinematic image, and the irreducibility of the gap dividing it from the space of the auditorium," Silverman argues, "filmic identification is almost definitionally excorporative" (Silverman 1996, 102). If incorporation describes the imaginary taking into the body of something external, and blending with it until it is indistinguishable from the original, then excorporation suggests the opposite: the splitting off of an imaginary original body, or a part thereof, and the attribution of, or blending with, an external other. In the cinema, this process is integral to the modes of idealization upon which our pleasures depend. As Silverman puts it, "the identification which always accompanies an idealization . . . necessarily follows an excorporative logic" (ibid., 96). Since cinematic projection conveniently fuses the technical and the psychic, perhaps excorporation might be thought of as the bodily equivalent to projection. Anticipating biological doubling, the cinema offers lifelike imitations that prepare the ground for the excorporations that science now makes available. Read psychoanalytically, the clone figure as identical copy emerges out of unconscious processes of subjectivity founded upon mirroring, doubling, and splitting. Following this line of thought, we might argue that the flickering images of life on the screen have prepared us for the prospect of our own biological doubling. Both generate uncanny phantasms of life at once real and artificial, organic and synthetic. The impossibility of seeing the difference between the original and the copy presents the subject with a compelling psychic scenario that is both utterly familiar within its own internal processes of identity formation and, despite that familiarity, strangely ungraspable as biological prospect.

Writing of just such a convergence between the arrival of the clone as literalization of the same as identical copy and the unconscious desires and identifications of the subject, the psychoanalyst Adam Phillips suggests that the "existence of the identical" stops a whole "range of political and psychological vocabularies . . . in their tracks" (Phillips 1998, 88–89). In so doing, we might say that it generates a theoretical double take. Drawing on two of his child patients' use of the clone in their narrations (see chapter 3), he suggests that perhaps, in representing the end of the continuing need "for two sexes in the task of reproduction," the clone figure "in one fell swoop" promises a "cure for sexuality and for difference . . . the sources of unbearable conflict" (ibid., 89). Phillips sees the present moment as the time "when the same has, as it were, been literalized as the identical . . . [and] seems to be a final solution to the problem of otherness" (ibid., 88–89). Cloning gives us the chance, he writes, to wonder "what's wrong with being apparently or exactly the same as someone else," and to wonder not only why we may be "frightened of being like other people—what is it we imagine we lose in the process" but also "why we may be frightened of other people being just like us" (ibid., 92). Perhaps, he suggests, the repudiation of the other as clone is indicative of a fear some people have of "not being a clone, of not being identical to someone, or identical to someone else's wishes for oneself" (ibid.). It is as if "we can only work out what or who we are like from the foundational belief—the unconscious assumption— that there is someone else that we are exactly like" (ibid.). In the end, Phillips argues, "cloning is used to get around history, as though in the total fantasy of cloning, history as difference is abolished" (ibid., 94). He concludes that perhaps even real cloning cannot satisfy a relentless wish for "absolute identity" that conceals "a profound doubt about our being the same as anything" (ibid.). To begin where Phillips ends, we might explore the relationship between identity and sameness, and between identity and what is not identical. According to Phillips, our ambivalence toward the figure of the clone rests on an apparent wish that masks another more troubling one. Perhaps that deeper fear lies in the elusiveness of relationality: that identity is not identical from the beginning.

In the clone figure, we confront the prospect of being unknowable both to ourselves and to others. The genetic imaginary is haunted by anxieties about the shifting ground of artificial identities, where absolute distinctions between originals and copies elude us, and where surface appearance seems to conceal secret truths. Mimicry and masquerade are transformed into experimental techniques of excorporative embodiment, and the subject's capacity for misrecognition extends ancient fears into fantasies of future techno-scientific possibilities.

Part of the disturbance presented by techno-scientific interference in genetic processes explored throughout this book is the prospect of the invisibility of genetic doubling—the troubling notion that genetic copying could be hidden in a body that appears perfectly normal from the outside. Just as the clone as biological double presents the ultimate vision of psychic impossibility, so the single embodiment of the consequences of genetic engineering realizes fears of duplicity, as the visual deceptions of genetic disguise and mistaken identity are tested.

In the event of genetic imposture or inadvertent Oedipal transgression, the double take extends into an uncanny sense of what we might call misembodiment: the feeling that the body encountered is not what it seems, or not what it should be (that there has been a loss of its bio-aura; see chapter 7). A most perplexing prospect is the notion that the clone figure might appear not as the double but as the singular misembodiment of genetic manipulation. The improvised and impossible kinships that could result require not just a second look but an extended epistemological double take.

The scopic regimes of the biological double as the literalization of the same as identical copy disturb the promise of clearer vision that genetic knowledge represents. The fissures between legibility, intelligibility, and visibility here return us to the desire for visual certainty that has been a concern throughout this book. Whereas visibility suggests there is something to be seen or to render visible, visuality suggests a seeing that is active and imaginative but perhaps less predictable. Visuality refers to a kind of "mental visibility," to "the state or quality of being visual to the mind," moving us away from fantasies of visible evidence or presence and toward a conceptualization of the genetic imaginary as a visualizing process through which subjects are constituted, and desires and identifications are inflected with unconscious preoccupations and repetitions.

The scenarios of the genetic imaginary repeat a desire for an impossible visualization that extends previous technological disturbances to the cellular body. In *Atomic Light (Shadow Optics)*, Akira Lippit deploys Jacques Derrida's notion of the "avisual" to conceptualize the convergence of technologies that transformed the "conditions of visuality as such" (Lippit 2005, 30). Lippit writes: "X-rays and cinema, along with the techniques of psychoanalysis, established in 1895 new technologies for visualizing the inside, for imagining interiority" (ibid.). For him, "this reconfiguration of the inside and out, surface and depth, visuality and avisuality" produced "the formation of a secret visuality . . . a new mode of avisuality" (ibid.). Accompanying these new phenomenologies, he claims, was a new mode of subjectivity which introduced "an alien you, secret and distant in its proximity to you . . . a secret visuality of you as another, of you who is here, in this image, pictured in this frame of a singular interiority, elsewhere and other" (ibid.). Lippit's discussion of what remains hidden in the atomic light of the twentieth century points to the inseparability of techniques of envisioning from perceptual modes of subjectivity. But the avisuality of genetically engineered and cloned bodies cannot quite be captured by Derrida's two formulations of the invisible, draw upon by Lippit. Suggestive as they are, neither the visible invisible (that which is in the order of the visible but is kept secret or out of sight, such as an organ in the body or something hidden by clothing), nor the absolute invisible (that which falls outside the register of sight, residing in another sense, for example) works in this context. Like the radioactivity of which Lippit writes, genetic information is invisible to the naked eye but can be tracked through visualizing techniques available to scientists. Unlike radioactivity, its avisuality contradicts our sense that we should be able to see the difference between the original and the copy (or that we can see someone's genes in their shared physical characteristics with others). Intimately tied to a language of observable physical resemblance, genetic discourse promises to make hidden biological truths appear on the surface of the body, as it translates the secrets of life into evident genealogies.

What Alys Eve Weinbaum calls "the blinding light of genetic reason" (Weinbaum 2008, 208) does not easily translate into visualizable knowledge. Genetic avisuality troubles an imaginary driven by the fantasy of a revelatory science. The gap between the precision of reading someone's genes technically (of being able to prove someone's presence at a crime scene from a minute fragment of skin, for example) and the elusiveness of the visual evidence of genetic artifice produces both the desire for certainty and the fear of its impossibility. The reconfiguration of visuality and subjectivity in the genetic imaginary takes Lippit's "secret other within" (2005, 31), born of late-nineteenth-century technologies, and translates it into the uncanny prospect of biological duplication. Now, with the advent of cloning, the "identical other without" of genetic excorporation haunts our sense of entitlement to unique individuality. Externalizing those secret others within through genetic literalization, the biological copy (which turns the same into the identical other) now faces us with an unseemly exposure. The discrepancy between the literalizing promise of genetic interference and the problem of visualization lingers, for even as cloning might instantiate an image of the self as other, the visual distinction between copy and original threatens to remain indiscernible.

The ontology of the visual image as already imitative mocks our desire for it to act as a place of certainty. The longing to see oneself in the gaze of the other (which, psychoanalytically speaking, will always be returned as misrecognition) is transformed into a fantasy of sameness in difference embodied in the biological copy. These modes of incorporation cut across cinematic and cloning techniques. They are matched by their opposite, excorporation—an externalization of the self in the other that also unites psychic and filmic processes with scientific ones. Perceptual disturbances to corporeal integrity and autonomy introduced by genetic experimentation thus interrupt the conventional flow of the identificatory pleasures of cinematic and scientific spectatorship.

The convergence of digitization and geneticization has led us to think about both the image and the body as comprising units of information.[3] By turning image and body into information, genetic and digital techniques allow us to reauthorize form and content. The possible extraction of the part from the whole disturbs the conventional flows of history, context, and genealogy. As temporalities are scrambled, kinships are warped, and new relationalities develop an unanchored life of their own, we face the prospect of what we might call post-Oedipal reproductions: modes of culture in which the traditional historical lineages of images and bodies lose their power to confer temporal authentication. Post-Oedipal reproduction refers to the dehistoricization of familial and kinship ties, the profound detraditionalization of heterosexual economies, and the loosening of the authenticating power of visual evidence in the culture of the copy.

The cinema sits at this convergence in a way that other media do not, as it struggles to imagine itself in a post-Oedipal future. The power of the cinematic image has been its potential to evoke our desire for repetition. Genre films have turned such pleasures into commodity recognition, as audiences have delighted in the difference in sameness that has become their trademark. But if the central myth of their narrative repetitions no longer holds sway over our psyches, and if the status of the image itself is in question, where lies the magic of the cinematic form that marked twentieth-century consumer cultures so profoundly? Does the prospect of the biological end of Oedipus also signal the death of the cinema? The disturbance to the power of the Oedipal narrative posed by genetic interference is not simply a transposition of interruption from biological to cultural form. Post-Oedipal reproduction troubles more than just this foundational narrative; it also transforms its modes of duplication, distribution, and circulation. Where there once were a

limited number of celluloid prints of films available for rental or purchase, since the development of video and then digital reproduction,
the availability of endless copies threatens previous regulatory mechanisms and legal and commercial interests. Old films can be transformed, remade, and remixed. Hollywood can be tampered with like
never before: Hitchcock's films can now be rescreened to last twenty-
four hours.[4] As information, bodies and images take on a life beyond
traditional modes of control and exhibition, their reproducibility and
mutability stretching into the domain of post-Oedipal interference and
its consequent ontological insecurities. Going far beyond the double
take, this is a world of revisioned multiplicities in which unoriginality is
not just expected but has become an art.

Alongside the proliferating practices of illegitimate and pirated copying and of downloading, ripping, and recombining afforded by computer and digital techniques are new modes of filmic spectatorship that
might intensify, rather than weaken, the cinema's powerful hold. According to Mulvey, the new viewing practices introduced first by video
and later by digital technologies have reframed the cinema's unique attribute in just such a way. The possibility of "returning to and repeating
a specific film fragment" afforded by postcelluloid media necessarily
means "interrupting the flow of film, delaying its progress, and in the
process, discovering the cinema's complex relation to time" (Mulvey
2006, 8). Photography shares with the cinema that sense of preserving
past time, but while the single image of the photograph "relates exclusively to its moment of registration," film has "an aesthetic structure
that (almost always) has a temporal dynamic imposed on it ultimately
by editing" (ibid., 13). It is this particular combination of the still and
the moving image and of the now-ness and then-ness that makes the
cinema so uniquely compelling. Although this tension between the still
frame and the moving image has arguably always been present in the
history of film, what Mulvey calls "delayed cinema" now reveals a new
perception of the "preserved time" that the medium presents: "in the
aesthetics of delay, the cinema's protean nature finds visibility" (ibid.,
12). In the "transitional period of 1995 that brought with it the metaphor
of the cinema's own death," we witnessed what Mulvey calls a new "ease
with which the cinema can be delayed" (ibid., 32). These new modes
of viewing have inaugurated a different relationship to the flow of the
images and to the invisible techniques that produced the illusions of

cinematic time, according to Mulvey. Bringing to the fore cinema's "hidden stillness," as Mulvey eloquently puts it, they generate an insatiable appetite for wanting to look again.[5]

This pause-and-look-again potential of new modes of cinematic spectatorship identified by Mulvey transforms the double take into a matter of technical desire—not the involuntary reflex of the split-second response, but the pleasure of controlling the flow and speed of images with the press of a button. Now we can stop a scene and gaze at the frozen shot for as long as we like; we can repeat a favorite moment as many times as we wish and return to those that contain crucial details we missed the first time around. Spectators with ideas above our station, we may feel omnipotent over the flow and sequence of the film. We seem to be in command of time. To repeat the past, freeze the present, halt time, fast-forward, predict the future—all of this now seems possible.

However pleasing such technologies of image control might seem to us, Mulvey suggests that these new modes of spectatorship reveal something disturbing. In these postcinematic haltings, she argues, we find the deathly haunting of the still image that reminds us of the cinema's inanimate origins. Accompanying such practices is "an amorphous . . . intangible difficulty [that] arises out of the presence of preserved time"; in the availability of old cinema through new technology, "a new kind of ontology" of "ambivalence, impurity and uncertainty" emerges (ibid., 11–12). Bringing to the surface the secret stillness of cinema's animating capacity, these new modes of spectatorship are accompanied by a troubling ontological insecurity. Mulvey cautions that even the freeze-frame of video and digital media, which seems to fulfill the desire to know through seeing, leaves a lingering uncertainty. Digital technology, she writes, "enables a spectator to still a film in a way that evokes a ghostly presence of the celluloid frame. Technically this is an anachronism" (ibid., 26). In intensifying the eerie quality of the cinema's animating capacity to make still images move as lifelike formations, making the image pause reiterates how "the blurred boundaries between the living and the not-living touch on unconscious anxieties that then circulate as fascination as well as fear in the cultures of the uncanny" (ibid., 32). Rather than delivering the reassurance that the omnipotence of such a halting of time and image might afford, the powers conferred by these technologies generate new anxiety instead. If the double take is driven by the expectation that a second look will do the trick, that the perceived discrepancy will be resolved through vision, here the pause-and-look-again potential disturbs the temporal certainties and continuities to which we are so attached.

It is this blurring of the boundary between life and death and between animate and inanimate that cuts across the psychic disturbances of cinematic and cloning techniques. Here the uncanny of the filmic image that returns us to our more ancient fears (even as it promises to resolve them through its pleasing idealizations) foreshadows the double take precipitated by the figure of the clone, whose artificial vitalities similarly lie between the familiar and the unfamiliar. In these instances of the technological uncanny, the maternal body that has been disposed of returns and demands that the subject take another look. The temporal disturbances of the post-Oedipal reproductions of genetic interference generate similar ontological insecurities. In both cases, fantasies of freezing time, returning to a lost past, and controlling the future circulate as parts are transposed, old and new remixed, and historical contexts or the flow of genealogies reinvented.

The proposition with which I end this book is that the genetic imaginary struggles to assimilate the prospect of technical interference in ways that have a prehistory in the cinema. If perception literally means to take in thoroughly, in attempting to take this form of uncanny life thoroughly, we stumble across the tracks of haunted territory. Cinematic life has compelled and disturbed us in ways that now surface in uncanny forms more clearly through the pause-and-look-again techniques of the postcelluloid image culture. The life of the gene is animated in laboratory techniques that signal a mode of post-Oedipal reproduction, giving biological form to the shock of the new of photographic and then cinematic reproduction so famously discussed by Benjamin (1968 [1936]). The cultural double take is a response to uncanny temporalities that can only fail to deliver the promised clarification of the second look. Interferences in the familiar visual and corporeal temporalities of the sequencing and flow of Oedipal reproductions are accompanied by the worrying sense that more vision might offer less certainty. This disturbance is aggravated rather than calmed by vision, as the repeated visualizations reiterate its elusiveness.

To posit a life of the gene in the cinema is to play upon the particularity of the medium's seductive power: to animate in the present something already lost to the past within traditional cinematic form. As the inanimate frames of film are brought to life on the screen, the projector reconciles the opposition between stillness and movement. Many film theorists have argued that death is never far away in the tension between the inanimate and the animate that generates what Mulvey calls the "entrancing illusions" that fill the cinema screen, images that "come to be more redolent of death than life" (Mulvey 2006, 11). Hidden in this movement is a haunting stillness that may or may not surface but that nevertheless "provides the cinema with its secret" (ibid., 67). Finding the motor of narrative movement in the cinema image's own trajectory, Mulvey suggests that "death as a trope that embodies narrative's stillness, its return to an inanimate form, extends to the cinema"; it is, she writes, "as though the still frame's association with death fuses into the death of the story, as though the beautiful automaton was to wind down into its inanimate, uncanny form" (ibid., 70). As a technology that animates images but is haunted by death, the cinema provides us with the ambivalent pleasure of revisiting the unthinkable boundary between the two.

The cinema is the only medium that is performative of the uncanny of the genetic imaginary, that fantasy space wherein the animations of techno-science undo deathly finalities, promising to return us to the familiar as they invent new modes of misembodiment. To the extent that film form has carved out the tracks of an uncanny doubling, the clone is a profoundly cinematic figure that enters our imaginary landscapes as the double we are sure we have met before. Reminding us of our previous excorporative identifications in the cinema, the clone invests the double take with a sense of déjà vu. The desire to look again, because we cannot tolerate our perception without a second look, is overlaid with that uncomfortable feeling that something (or someone or somewhere) unfamiliar is nevertheless already known to us. The cinematic life of the gene requires that we look again, as it transports us back to a time and place where the new and unassimilable is all too familiar.

The cinema occupies pride of place in the genetic imaginary. No other medium or practice shares the animating project of genetic science that, in promising to generate life, endlessly reminds us of the prospect of death. The genetic freeze frames that disavow the impossibility of halting life and delaying death prompt both the double take and a sense of déjà vu. The intensity of bewildering affect and the visual elusiveness that drive the genetic imaginary take their force from these new dimensions of strange familiarity. Like the illusions of the cinema that demand we invest in the present and presence of lost time, the seductions of a science that can reinvent its own temporalities leave us with the desire both to look away and to look again.

Notes

Introduction

1. *Gattaca* is discussed in full in chapter 5. I am grateful to Celia Roberts for discussions that led to this focus for the opening of my introduction.
2. See Franklin 2007 for a study of Dolly the sheep.
3. Elizabeth Grosz discusses the image of the Möbius strip in relation to theories of sexuality (Grosz 1995, 97).
4. There is a vast literature on visualization of the body's interior as part of the project of modernity—for example, Foucault 1973; Crary 1990; and Stafford 1991. For discussions of the gendering of this desire, see Jordanova and Hulme 1990; Bruno 1992; Shohat 1992b; Duden 1993; Braidotti 1994; Cartwright 1995; Marchessault and Sawchuk 2000; and van Dijck 2005.
5. See Foucault 1978; Duden 1993; Rose 1996; Haraway 1997; and Franklin 2000.
6. For a discussion of the relationships between these categories, see chapters 4 and 5. See also Doane 1991a.
7. Critical work on the genre of science-fiction film includes: Kuhn 1990 and 1999; Penley et al. 1991; Sobchack 1997; Telotte 1999 and 2001; Newman 2002; and Wood 2002.

8. For a discussion of film comedies concerned with genetic mixing, see Roof 1996b and 2007.

9. See van Dijck 1998 and Franklin 2000, 198.

10. See Landecker 2007.

11. My approach here draws upon the work of Franklin 2000, but differs from it in significant ways.

12. On gene fetishism, see also David Le Breton, who claims that "the fetishizing of DNA which . . . effectively eliminates concrete life, is one of the most perturbing sociological phenomena in the contemporary world" (2004, 1). For a summary of poststructuralist writing on genetics that extends into the idea of the "genetic imaginary," see Egorova, Edgar, and Pattison 2006.

13. For important work using psychoanalytic film theory to analyze scientific constructions of the body, see Bruno 1992 and Doane 1990.

14. For accounts of the historical emergence of these theoretical traditions, see Coward and Ellis 1977; Hall, Dobson, Lowe, and Willis 1980; and Hollway, Venn, Walkerdine, and Henriques 1998.

15. For a more detailed timeline of cloning, see Haran et al. 2007, 186–89.

16. See Haran et al. 2007 for an analysis of this context.

17. For a discussion of the convergence between the digital and the genetic, see especially Mackenzie 2002 and Thacker 2004.

18. A survey in Haran et al. 2007 lists 60 cloning films that were released between 1995 and 2006, compared with only 12 in the previous ten years.

19. See also Nottingham 2000. For a broad account of representations of cloning in film, literature, art, print journalism, and advertising, see Haran et al. 2007.

20. For feminist rereadings of kinship theory, see Butler 2000 and 2002.

21. I am grateful to Lesley Stern for discussions on this point at my March 2006 talk in the Department of Communications of the University of California at San Diego.

22. See especially Bernardi 1996 and 2001; Foster 2003; Vera and Gordon 2003; and Dyer 2004.

1. The Hell of the Same

1. One exception to this is Adam Phillips 1998 whose writing on cloning is discussed in Chapter 3 and in the Afterword.

2. For analyses of cultural constructions of the gene, see Nelkin and Lindee 1995; Keller 1995, 2000; Rothman 1998; Turney 1998; van Dijck 1998; Condit 1999; Franklin 2000; and Roof 2007.

3. As I discuss in the introduction, Donna Haraway 1997; Sarah Kember 1998; and Judith Roof 1996b also address this question.

4. I am grateful to Andrew Quick for drawing my attention to some of this work and for ongoing discussions about its significance.

5. For a discussion of the uses of rhetoric within theory more generally, see Pearce 2003.

6. Previous versions of these articles have been reproduced in a number of forms which vary only slightly; see Baudrillard 1993 and 1997.

7. For an excellent account of the traces of colonial constructions of primitive otherness in Freudian psychoanalysis, see Khanna 2003.

8. For a cultural analysis of Dolly the sheep, see Franklin 2007.

9. For a fuller exposition of this argument, see Butler 1990.

10. Within the French context, opponents of civil unions for same-sex couples have mobilised rigid structuralist and psychoanalytic theoretical models of cultural and biological reproduction to challenge the proposed legal recognition of lesbian and gay partnerships, and, most especially rights to parenthood and adoption where "the figure of the child of nonheterosexual parents becomes a cathected site for anxieties about cultural purity and cultural transmission" (Butler 2002, 23).

11. For an important contribution to rethinking traditional studies of kinship in anthropology which Judith Butler's argument also draws upon, see Franklin and McKinnon 2001.

2. She Is Not Herself

1. Battaglia's excellent essay on human replication in the popular cinema focuses particularly on the film *Multiplicity* (Harold Ramis, 1996).

2. Battaglia (2001, 496) offers an analysis of replicants and clones in the cinema "as corporealizations of the capacity of the supplement to destabilize the social paradigms and self-knowledge of their creators." In this definition of the "supplement," Battaglia borrows from Derrida.

3. For a more general discussion of the "proliferation of sameness" in relation to the shift from reproduction to replication in the context of cloning, see Squier 1999.

4. The cultural significance of the kinds of monstrous bodies that populate the *Alien* films has been widely debated. See in particular Kavanaugh 1990; J. Newton 1990; Penley 1991; Hurley 1995; E. Hills 1999; and Gibson 2001.

5. For a discussion of the cultural significance of the pixel, see Franklin, Lury, and Stacey (2000, 60–66).

6. Cooper's article presents a fascinating discussion of the history of the scientific study of teratomas in relation to the development of contemporary stem-cell research, looking in particular at the history of conceptualization of the normal and the pathological in teratology and teratogeny.

7. In a woman, a teratoma may imitate pregnancy insofar as it has the capacity to produce the cells of all the organs in the body. Thus, unlike many tumors, teratomas have the fascinating and yet grotesque characteristic of fetal resemblance. This is why teratomas are also called monstrous births. For a more detailed discussion of teratomas, see Stacey 1997.

8. As Braidotti (1994) has shown, teratology, or the scientific study of monsters, was the forerunner of contemporary embryology.

9. This argument is developed in Silverman's (1991) reading of *Blade Runner*. For a comprehensive analysis of this film, see Bukatman 1997.

10. For important analyses of popular cultural constructions of the gene, see Nelkin and Lindee 1995; Turney 1998; and van Dijck 1998.

11. See Franklin 2000 on the genetic imginary. For a discussion of images of the "helping hand" of science in new genetic interventions, see Franklin 2002.

12. Battaglia (2001, 507) points out that the advertising slogan of the Tyrell Corporation (the company that produced the replicants in *Blade Runner*) is "More Human than the Humans." On the human-replicant relation, see also Silverman 1991, 113–15.

13. For a discussion of the incestuous character of the complex forms of desire between Ripley and the various aliens in the film, see Gibson 2001, 49.

14. For a discussion of the commoditization of memory in science-fiction films, see Bukatman 1997, 77–80.

15. Constable here draws upon Christine Battersby's (1998) reworking of Luce Irigaray's theories of feminine subjectivity through the notion that, because of her potential to give birth to another, the female subject might be used to provide a new model of interrelatedness: as being more than one and less than two through the connection to another's bodily formation.

16. For further discussion of action heroines, see Clover 1992; Tasker 1993 and 1998; Jennings 1995; and Holmlund 2002.

17. For an analysis of Michael Jackson in this video, see Mercer 1986.

18. For an analysis of androgyny and sexuality in science-fiction films, see Bergstrom 1991.

19. For a discussion of gay male cloning cultures, see Lauritsen 1993.

20. For a discussion of the sexualized use of hands in lesbian films, see Merck 2000.

21. For an analysis of the posthuman, see Hayles 1999. I am grateful to Michael Lundin for letting me read his unpublished paper on *Alien: Resurrection* as a presentation of posthuman embodied subjectivity (2002). That work is drawn from his Ph.D. dissertation in progress (Lundin n.d.).

22. For a discussion of the remaking of "life itself" through genetics, see Franklin 2000.

23. I am grateful to Imogen Tyler for discussions on these points. For an anal-

ysis of the disruptions to psychoanalytic categories produced by pregnant embodiment, see I. Tyler 2001.

3. Screening the Gene

1. For a lucid account of dialogic theory, see Pearce 1994.
2. See chapter 1 for a discussion of important analyses of cultural constructions of the gene, such as Nelkin and Lindee 1995; Turney 1998; van Dijck 1998; and Franklin 2000.
3. Van Dijck (1998, 38–44) argues that James Watson's *The Double Helix* (1968) performs the same structure as the model of the DNA molecule.
4. See Roof 1996b and 2007.
5. I am grateful to Charlotte Brunsdon for suggesting this term.
6. I am grateful to Maureen McNeil and to Charlotte Brunsdon for discussions of this point.
7. For a discussion of the emergence of film narratives concerned with the ethics of preemptive logic, see Weber 2006.
8. For an excellent analysis of the multiplicity of screens in postmodern culture, see Friedberg 1993.
9. Braidotti offers important analyses of the current fascination with mutational bodies in Hollywood cinema (2002) and of new ways of thinking about the changing dynamics of life and death (2006).
10. The train which transports Sil from Utah to Los Angeles becomes a spectacular monstrous body during her metamorphosis; it was designed by H. R. Geiger, who also designed the special effects for *Alien* and *Alien: Resurrection*.
11. In this subgenre, M. Hills (2005) includes *Alien: Resurrection, Deep Blue Sea, The Island of Dr. Moreau, Jurassic Park, Mimic* (Guillermo del Toro, 1997), *The Relic* (Peter Hymans, 1997), and *Species*, as well as the sequels *Mimic 2* (Jean de Segonzac, 2001) and *Species 2* (Peter Medak, 1998).
12. For discussions of how fetishism is embedded within the apparatus of the cinema, see Mulvey 1989 and 1996.

4. Cloning as Biomimicry

1. For theories of feminine embodiments as more than one and less than two, see Kristeva 1982, Irigaray 1985, and Battersby 1998.
2. For a feminist discussion of uses of the word "narcissism" more generally as a cultural category, see C. Tyler 2003.
3. See Latour and Weibel 2005.

4. On double mimesis, see Silverman 1996, especially 59–98. On the mimesis of savagery, see Low 1996.

5. For fuller discussions of both surface and depth and of passing, see Butler 1990 and 1993, respectively.

5. Genetic Impersonation

1. Although I used a discussion of a scene from *Gattaca* to open the book, I shall reiterate some synoptic points about the film here in a slightly different form, for readers who have turned straight to this chapter.

2. The touchability of the image here is reminiscent of what Marks (2000) has called a "haptic" quality.

3. The relationships among technology, surveillance, and knowledge is a familiar theme in science fiction, as in *Minority Report* (Steven Spielberg, 2002). For cultural analyses of science-fiction films more generally, see Kuhn 1990; Penley et al. 1991; Sobchack 1997; Kuhn 1999; Telotte 1999; and Wood 2002.

4. As Silverman (1991) has argued in relation to *Blade Runner* (Ridley Scott, 1982), slavery separated from race provides a fertile fantasy for the threat of replication in science-fiction films.

5. For a discussion of alternative kinship modes within lesbian, gay, and queer communities, see Weston 1991.

6. I am grateful to Lauren Berlant for suggesting the word "improvisation" here.

7. For an analysis of the blood tie and kinship theory, see Franklin and McKinnon 2001.

8. I am indebted to the postmodern genealogies reading group at Lancaster University for discussions about queer kinship and shared bodily substances.

9. The suspense about who is authentic and who is artificial is endlessly rehearsed in science-fiction films, perhaps most famously in *Blade Runner*.

10. There is now a vast literature that builds on the earlier work of Dyer (1982) and Neale (1983) on the construction of masculinity as spectacle. For more recent analyses, see Cohan and Hark 1993; Kirkham and Thumim 1993; Tasker 1993; Lehman 2001; and Holmlund 2002.

11. Johnston (1975) first used the word "masquerade" to analyze the place of woman in film. For a detailed genealogy of the uses of masquerade as a psychoanalytic term in film theory during the 1970s and 1980s, see Fletcher 1988.

12. In her famous 1929 essay, "Womanliness as a Masquerade," Riviere discusses the case of a female academic who exaggerated her femininity fol-

lowing successful public-speaking engagements as a way to contradict her theft of masculine authority (Riviere 1986).

13. For a detailed discussion of masculinity and masquerade, see also Studlar 1996.

14. For discussions of the display of the masculine body through imitation and disguise, see Cohan 1992, 397, and Holmlund 1993.

15. For a discussion of "female masculinity" in film, see Halberstam 1998.

16. For a discussion of "life itself" in the context of genetic engineering, see Franklin 2000.

17. Braidotti (2002, 222–63) offers a critique of how fantasies of reproductive technologies in science-fiction films have produced the reinvention of a paternity that allows the exclusion of women from reproduction.

18. In his work on Hollywood stars, Shingler (1995, 192) has pointed toward a theoretical model which might combine psychoanalytic theories of masquerade with Butler's rereading of gender impersonation through drag.

19. For Cohan, Butler's (1990, 25) insistence on gender as "performative—that is, constituting the identity it is purported to be" allows for the considering of masculinity as masquerade in film, not as a mask but as a "persona" (Cohan 1992, 398).

20. This scene echoes Scottie's famous attempt to overcome his anxiety and climb the steps of the church tower in *Vertigo* (Alfred Hitchcock, 1958). See Modleski 1988.

21. The investment of Jerome with a masculinity so desirable that others want to become him has been a recurrent dimension of Jude Law's star image. See, for example, *The Talented Mr. Ripley* (Anthony Minghella, 1999) and Straayer 2001. In both *eXistenZ* (David Cronenberg, 1999) and *Artificial Intelligence* (Steven Spielberg, 2001), Law is positioned as the object of desire; he embodies the ideals of masculine perfection, but his relationship to artificiality places him at one remove from such an embodiment.

22. I am grateful to Lauren Berlant for discussions of this mode of kinship.

23. Along with lesbian and gay passing, there are other forms of passing that are obviously connoted here: the rejection of Vincent from education and employment in *Gattaca* is reminiscent of the exclusion of black people from schools and jobs in the United States prior to the civil-rights movement. For a discussion of racial passing, see Young 1996 and Ahmed 1999.

24. The triangulated dynamics of homosocial desire, normally opposed to homosexual desire, are the subject of Sedgwick's (1985) analysis of such an ascription. For a discussion of these triangular dynamics in the novel *The Talented Mr. Ripley*, see Straayer 2001, 115–32.

25. Thanks to Amelia Jones for her insightful comments on our underlying fears about cloning.

26. For an analysis of the concept of panhumanity in global culture, see Franklin, Lury, and Stacey 2000, 37–42.

6. Uncanny Architectures of Intimacy

1. For feminist work on the new kinship implications of genetic engineering, see Franklin and McKinnon 2001.
2. See Kember 2003 and de Lauretis 2003.
3. See Sobchack 1997; Kuhn 1999; and Telotte 2001.
4. I am grateful to Lynne Pearce for suggesting this term after reading an earlier version of this chapter. For a discussion of the temporalities of postmodern romance, see Pearce 2007.
5. Here I am referring to Freud's notion of the primal scene, developed through his reading of the famous Wolf Man dream. See Freud (1918) 2002.
6. For a discussion of the sociological notion of disembedding, see Giddens 1990. For an analysis of the notion of life itself, see Foucault 1979; Duden 1993; and Franklin 2000.
7. In the director's comments accompanying the DVD of *Code 46*, Winterbottom stresses the importance of the use of 35-mm film and on-location shooting to his vision for the film. I am grateful to Nitin Govil for raising this point.
8. For a critique of the essentializing discourse of techno-orientalism, see Orbaugh 2005.
9. This phrase was used by Jacqueline Orr in conversation with me about the culture of biogenetics.
10. For a discussion of the monstrous double in science fiction, see Telotte 1990. For a discussion of science fiction as an uncanny text, see Telotte 2001.
11. Michael Winterbottom's interest in questions of migration, invisibility, and transnational flows of bodies, images, and information is also evident in his previous film *In This World* (2002).
12. The concept of panhumanity was originally developed in relation to a critique of the new universalisms of globalizing cultures. See Franklin, Lury, and Stacey 2000.
13. For a discussion of intimacy and the multicultural subject, see Fortier 2008.
14. I am grateful to Cynthia Weber for discussions about this aspect of the film.
15. For a discussion of the changing meaning of genealogy and temporality, see Franklin and McKinnon 2001.
16. In the *Elementary Structures of Kinship* (1969) Claude Lévi-Strauss argues

that the universal belongs to nature and that which is variable from one social structure to another belongs to culture and yet granted that the incest taboo reached into both realms. It is this contradiction which Derrida is in dialogue with here. I am grateful to Julie Crawford and Liza Yukins for raising this question in discussion about the film.

17. These possible multiple relationalities bring to mind the famous line of Evelyn Mulwray (Faye Dunaway) in *Chinatown* (Roman Polanski, 1974), which reveals the kinship transgressions resulting from her father's sexual abuse: "She's my daughter! She's my sister! She's my daughter! My sister, my daughter. She's my sister *and* my daughter."

18. The notion of prosthetic memory (of memory's extraction or implantation) is explored in a number of films as well as in *Code 46*—see, for example, *Eternal Sunshine of the Spotless Mind* (Michel Gondry, 2004), and *Memento* (Christopher Nolan, 2000), and *Robocop* (Paul Verhoeven, 1987). The question of memory as the marker of authentic humanness is also part of *Blade Runner*, to which *Code 46* is so clearly indebted.

19. Kate O'Riordan (2006) draws attention to the predominant use of hands in this sex scene. See also Haran et al. 2007.

20. I am grateful to Peter Dickinson for reading an earlier draft of this chapter and pushing me to clarify this connection.

21. This equation is similar to staging of a conflict between the sexual and the political in *1984* (Michael Anderson, 1956).

22. Nostalgia is "melancholia caused by prolonged absence from one's home or country"(*OED*, n. 1).

23. I am grateful to Maureen McNeil for discussions about how to read this final scene.

7. Cut-and-Paste Bodies

1. This essay by Benjamin is "probably the most frequently and most intensively debated essay in the history of the academic humanities of the twentieth century" (Gumbrecht and Marrinan 2003, xiii), and it is hard to imagine what could possibly be left unsaid about it. The work has been thoroughly discussed not only within literature, film, and cultural studies, but also across sociology, political theory, and philosophy; these analyses need no further rehearsal here. My point rather is to bring this essay into the debates about genetic cultures—where its implications have yet to be debated (with the exception of Weinbaum 2008).

2. For a discussion of the analogy with painting here, see Mitchell 1992.

3. For an excellent discussion of visibility and indexicality, see Ostherr 2005.

8. Leading Across the In-Between

1. See Cubitt 1998 and 2004, Manovich 2001, and Willemen 2002.

2. See for example, Mulvey 1989 and 1996, Doane 1991c, Hart 1994, and Kaplan 1998.

3. For a fuller discussion of the masquerade debates, see chapter 4 and Doane 1991c.

4. See Dyer 2006 for a discussion of pastiche in the cinema.

5. For critiques of the shared desires of science and the cinema, see especially Bruno 1992 and Shohat 1992a.

6. Discovered by the French in 1799, the Rosetta Stone has the same text carved on it in three different languages. Comparative translations of the text led to the deciphering of ancient hieroglyphic writing.

7. As Marsha Kinder points out, the transsexual writer and activist Allucquère Rosanne Stone (aka Sandy Stone) is invoked in the film in the names Sandy and Rosetta Stone, and in the "film's thematics—Rosetta's multiple identities and Ruby's vampiric sexuality" (Kinder 2005, 177).

8. The director suggests on the DVD that the silk kimonos were selected for a practical reason: so that Tilda Swinton could quickly change from one character to another. Practicalities notwithstanding, the effect is one of Easternization.

9. For example, in relation to acoustic measurement, a transducer is the instrument which must transduce the input signals: "the dynamic range of a microphone is the range of levels of input signals which can usefully be transduced by the instrument." In neuroscience, light must be transduced into a neural message: "Rhodopsin, a visual pigment found in vertebrate retinal rods, transduces light into a neural message." In a 1977 usage, "carbon granules . . . do their delicate work of transducing sound waves into varying electric currents" (OED, transduce, definition 1), and from the 1950s on, in microbiology, transduction means "the transfer of genetic material from one cell to another by a virus or virus-like particle" (OED, transduction, definition 3).

10. See Kaplan 1980, Modleski 1988, Mulvey 1989 and 1996, Kuhn 1994, and Thornham 1997 and 1999.

11. See Gledhill 1991, Dyer 1998 and 2004, and Moseley 2005.

12. In his formulation of biomedia, Thacker draws on Grusin and Bolter's theory of the double logic of "remediation," in which they identify a contradictory imperative in the current cultural desire for "immediacy and hypermediacy," the desire both to "multiply its media and to erase all its traces of mediation" (Grusin and Bolter 1999, 5).

13. For example, *Chinatown* (Roman Polanski, 1974) is cited at two different points in the film: first, when Marinne talks about the complexity of her

relationship to Rosetta and mockingly slaps her own face from one side to the other as she says, "she's my mother, my sister, my mother, my sister" (paraphrasing Evelyn Mulwray [Faye Dunaway]); and second, when a roaming Band-Aid inexplicably appears (first on his nose, then on his chin and his cheek) on the investigator's face, referencing the one on the face of J. J. Gittes (Jack Nicholson).

14. The intertextual and citational aesthetic of the film extends from the cinema into contemporary art practice and beyond (not surprisingly, given that most of the director's work has been not in film but in new-media art). There is an extensive set of borrowings from popular culture, art, literature, and politics. For example, Eduardo Kac's green rabbit (Alba)—one of the icons of scientific and artistic collaboration known as GenArt—appears in the laboratory scene behind two scientists who are debating the ethics of genetic engineering in their laboratory; and the Edward Hopper-style bar scene where we witness Ruby's seduction techniques for the second time transforms artistic style into mise-en-scène (this connection is further played on when the investigator of bioterrorism has the same surname as the artist).

15. See http://agentruby.sfmoma.org (accessed 20 November 2009).

16. Claudia Springer first used the term "sexual interface." See Springer 1996.

17. Here Thacker's question "what is 'biomolecular affect'?" would belong to this more general desire to literalize affect. He writes: "We are used to thinking of affect and phenomenological experience generally in anthropomorphic terms," but "is there a phenomenology of molecular biology? Are there zones of affect specific to the molecular domain and irreducible to . . . anthropomorphicisms? What would such an analysis say concerning our common notions of embodied subjectivity?" (Thacker 2004, 31).

18. The controversial affective turn is widely debated across the humanities and social sciences: see Buchanan and Colebrook 2000, Ahmed 2004, Brennan 2004, Hemmings 2005, and Braidotti 2006. Angerer 2007 (the title of which translates as "The Desire for Affect") is currently available only in German.

9. Enacting the Gene

1. For a discussion of the monsters and marvels of science in a different, but related, area see Stacey 1997.

2. See Franklin 2000 as well as Foucault 1979 and Duden 1993 for a discussion of life itself.

3. The 1919 collage on paper is at the Institute for Foreign Cultural Relations in Stuttgart.

4. The concept of disembedding and reembedding is explored by Anthony Giddens (1990) in his sociological theory of modernity, *The Consequences of Modernity*.

5. Ontario Place is defined on its Web site as "a cultural, leisure and entertainment parkland . . . [which] extends throughout three man-made islands along the Lake Ontario waterfront . . . and opened in May 1971."

6. There are many excellent feminist cultural studies of techno-science. See for example Cartwright 1995; Treichler, Cartwright, and Penley 1998; Marchessault and Sawchuk 2000; Waldby 2000; and van Dijck 2005.

7. Examples include *The Incredible Shrinking Man* (Jack Arnold, 1957), *Honey, I Shrunk the Kids* (Joe Johnston, 1989), and *Toy Story* (John Lasseter, 1995), as well as Lewis Carroll's *Alice in Wonderland* (1993) and Jonathan Swift's *Gulliver's Travels* (2001).

8. For feminist analyses of Julia Kristeva's theory of abjection, which involves both repulsion and fascination, see Barbara Creed 1993 on "the monstrous feminine," and Rosi Braidotti 2002 on the teratological imagination.

9. For a historical account of the convergence between scientific and cinematic visualizing techniques, see Cartwright 1995.

10. One of the first analyses to make such connections between science and masculine national heroism was McNeil 1987.

11. I am grateful to Joy James for discussions of the current cultural fascination with the minute and the microscopic and for pointing out to me the emergence of what is referred to as "the CSI effect," through which a television program has been deemed to alter the ways in which forensic evidence is interpreted by juries. On the history of photographic techniques, see James 2003.

12. For a discussion of the idea of recombinant bodies and genetic science, see Steinberg 1997.

13. For a detailed description of computer collage, see Mitchell 1992, 162–89.

14. *Genetic Admiration* is distributed by Vtape: www.vtape.org (401 Richmond St. West, #452, Toronto, Ontario M5V 3A8, Canada).

Afterword

1. This phrase is taken from Mary Douglas's *Purity and Danger: An Analysis of Concepts of Pollution and Taboo* (1966).

2. The phrase "monsters and marvels" is borrowed from my own cultural study of cancer; see Stacey 1997.

3. For a discussion of this convergence, see van Dijck 1998. For the most recent work on genomic cultures, see Haran et al. 2008.

4. I am referring here to Douglas Gordon's 1993 video projection *24 Hour Psycho*, in which the film was slowed to last twenty-four hours.
5. The phrase "hidden stillness" is taken from the back cover of Mulvey's book. For a discussion of the tension between the photographic and the cinematic in relation to stillness and movement, see Silverman 1996.

Bibliography

Ahmed, Sara. 1999. "She'll Wake Up One of These Days and Find She's Turned into a Nigger: Passing through Hybridity." *Theory, Culture and Society* 16, no. 2: 87–106.

———. 2000. *Strange Encounters: Embodied Others in Post-Coloniality.* London: Routledge.

———. 2004. *The Cultural Politics of Emotion.* Edinburgh: Edinburgh University Press.

Angerer, Marie-Luise. 2000. "To Be the ONE—Männlichkeit als obsessionale Reihe." In I. Bieringer, W. Buchacher, and E. Forster, eds., *In Männlichkeit und Gewalt*, 91–95. Opladen, Germany: Leske und Budrich.

———. 2007. *Das Begehren nach dem Affekt.* Berlin: Diaphanes.

Anker, Suzanne, and Dorothy Nelkin. 2004. *The Molecular Gaze: Art in the Genetic Age.* Woodbury, N.Y.: Cold Spring Harbor Laboratory Press.

Appadurai, Arjun. 1996. *Modernity at Large: Cultural Dimensions of Globalization.* Minneapolis: University of Minnesota Press.

Augé, Marc. 1995. *Non-places: Introduction to an Anthropology of Supermodernity.* New York: Verso.

Austin, J. L. 1962. *How to Do Things with Words.* Oxford: Clarendon Press of Oxford University Press.

Baecker, Dirk. 2003. "The Unique Appearance of Distance." In Hans Ulrich

Gumbrecht and Michael Marrinan, eds., *Mapping Benjamin: The Work of Art in the Digital Age*, 9–23. Stanford, Calif.: Stanford University Press.

Barthes, Roland. 1981. *Camera Lucida: Reflections on Photography*. Translated by R. Howard. New York: Hill and Wang.

Battaglia, Debbora. 2001. "Multiplicities: An Anthropologist's Thoughts on Replicants and Clones in Popular Film." *Critical Inquiry* 27, no. 3: 493–514.

Battersby, Christine. 1998. *The Phenomenal Woman: Feminist Metaphysics and the Patterns of Identity*. Cambridge, U.K.: Polity.

Baudrillard, Jean. 1993. *The Transparency of Evil: Essays on Extreme Phenomena*. Translated by J. Benedict. London: Verso.

———. 1997. *Simulacra and Simulation*. Translated by S. F. Glaser. Ann Arbor: University of Michigan Press.

———. 2000. *The Vital Illusion*. Edited by Julia Witwer. New York: Columbia University Press.

———. 2002. *Screened Out*. Translated by Chris Turner. London: Verso.

Bellour, Raymond. 1996. "The Double Helix." In Timothy Druckery, ed., *Electronic Culture: Technology and Visual Representation*, 173–99. New York: Aperture.

Benjamin, Walter. 1999 [1936]. "The Work of Art in the Age of Mechanical Reproduction." *Illuminations: Essays and Reflections*, edited and with an introduction by Hannah Arendt and translated by Harry Zohn, 211–45. London: Pimlico.

———. 1978. *Reflections*. Translated by E. Jephcott. New York: Schocken.

Bergstrom, Janet. 1991. "Androids and Androgyny." In Constance Penley, Elisabeth Lyon, Lynn Spigel, and Janet Bergstrom, eds., *Close Encounters: Film, Feminism, and Science Fiction*, 33–62. Minneapolis: University of Minnesota Press.

Berlant, Lauren. 2008. *The Female Complaint: The Unfinished Business of Sentimentality in American Culture*. Durham, N.C.: Duke University Press.

Bernardi, Daniel. 1996. *The Birth of Whiteness: Race and the Emergence of U.S. Cinema*. New Brunswick, N.J.: Rutgers University Press.

———. 2001. *Classic Hollywood, Classic Whiteness*. Minneapolis: University of Minnesota Press.

Bersani, Leo, and Ulysse Dutoit. 2004. *Forms of Being: Cinema, Aesthetics, Subjectivity*. London: British Film Institute.

Bhabha, Homi K. 2005. *The Location of Culture*. New York: Routledge.

Bolter, Jay David, and Diane Gromala. 2003. *Windows and Mirrors: Interaction Design, Digital Art, and the Myth of Transparency*. Cambridge: MIT Press.

Boss, Pete. 1986. "Vile Bodies and Bad Medicine." *Screen* 27, no. 1: 14–24.

Braidotti, Rosi. 1994. *Nomadic Subjects: Embodiment and Sexual Difference in Contemporary Feminist Thought*. New York: Columbia University Press.

———. 2002. *Metamorphoses: Towards a Materialist Theory of Becoming*. Cambridge, U.K.: Polity.

———. 2006. *Transpositions: On Nomadic Ethics*. Cambridge, U.K.: Polity.

Brennan, Teresa. 2004. *The Transmission of Affect*. Ithaca, N.Y.: Cornell University Press.

Brophy, Philip. 1986. "Horrality—the Textuality of Contemporary Horror Films." *Screen* 27, no. 1: 2–13.

Bruno, Giuliana. 1987. "Ramble City: Postmodernism and *Blade Runner*." *October* 41, summer: 61–74.

———. 1992. "Spectatorial Embodiments: Anatomies of the Visible and the Female Bodyscape." *Camera Obscura* 28: 239–62.

Brunsdon, Charlotte. 2006. "A Fine and Private Place: The Cinematic Spaces of the London Underground." *Screen* 47, no. 1: 1–17.

Buchanan, Ian, and Claire Colebrook, eds. 2000. *Deleuze and Feminist Theory*. Edinburgh: University of Edinburgh Press.

Bukatman, Scott. 1997. *Blade Runner*. London: British Film Institute.

Butler, Judith. 1990. *Gender Trouble: Feminism and the Subversion of Identity*. London: Routledge.

———. 1991. "Imitation and Gender Insubordination." In Diana Fuss, ed., *Inside/Out: Lesbian Theories, Gay Theories*, 13–31. London: Routledge.

———. 1993. *Bodies That Matter: On the Discursive Limits of Sex*. New York: Routledge.

———. 1997. *The Psychic Life of Power: Theories in Subjection*. Stanford, Calif.: Stanford University Press.

———. 2000. *Antigone's Claim: Kinship between Life and Death*. New York: Columbia University Press.

———. 2002. "Is Kinship Always Already Heterosexual?" *differences: A Journal of Feminist Cultural Studies* 13, no. 1: 14–44.

———. 2005. *Giving an Account of Oneself*. New York: Fordham University Press.

Caillois, Roger. 2003 [1935]. "Mimicry and Legendary Psychasthenia." In Claudine Frank, ed., *The Edge of Surrealism: A Roger Caillois Reader*, 90–103. Durham, N.C.: Duke University Press.

Carroll, Lewis. 1993. *Alice in Wonderland*. Ware, England: Wordsworth Editions.

Cartwright, Lisa. 1995. *Screening the Body: Tracing the Visual Cultures of Biomedicine*. Minneapolis: University of Minnesota Press.

Castañeda, Claudia. 2003. *Figurations: Child, Bodies, Worlds*. Durham, N.C.: Duke University Press.

Clover, Carol. 1992. *Men, Women and Chainsaws: Gender in the Modern Horror Film*. London: British Film Institute.

Cohan, Steve. 1992. "Cary Grant in the Fifties: Indiscretions of the Bachelor's Masquerade." *Screen* 33, no. 4: 394–412.

Cohan, Steve, and Ina Rae Hark, eds. 1993. *Screening the Male: Exploring Masculinity in Hollywood Cinema*. London: Routledge.

Coleman, Beth. 2006. "Mr. Softee Takes Command: Morphological Soft Machines." *Can We Fall In Love with a Machine?* Wood Street Gallery, Pittsburgh, Pa.

Condit, Celeste Michele. 1999. *The Meanings of the Gene: Public Debates about Human Heredity.* Madison: University of Wisconsin Press.

Constable, Catherine. 1999. "Becoming the Monster's Mother: Morphologies of Identity in the *Alien* Series." In Annette Kuhn (ed.), *Alien Zone II: The Spaces of Science Fiction Cinema,* 173–202. London: Verso.

Cooper, Melinda. 2004. "Regenerative Medicine: Stem Cells and the Science of Monstrosity." *Medical Humanities* 30: 12–22.

Copjec, Joan. 1994. *Read My Desire: Lacan against the Historicists.* Cambridge: MIT Press.

Coward, Rosalind, and John Ellis. 1977. *Language and Materialism: Developments in Semiology and the Theory of the Subject.* London: Routledge.

Crary, Jonathan. 1990. *Techniques of the Observer: On Vision and Modernity in the Nineteenth Century.* Cambridge: MIT Press.

Creed, Barbara. 1993. *The Monstrous Feminine: Film, Feminism and Psychoanalysis.* London: Routledge.

Cubitt, Sean. 1998. *Digital Aesthetics.* London: Sage.

———. 2004. *The Cinema Effect.* Cambridge: MIT Press.

de Certeau, Michel. 1984. *The Practice of Everyday Life.* Translated by S. Rendall. Berkeley: University of California Press.

de Lauretis, Teresa. 2003. "Becoming Inorganic." *Critical Inquiry* 29: 547–70.

Deleuze, Gilles. 1992. "Postscript on the Societies of Control." *October* 59, winter: 3–7.

Derrida, Jacques. 1981 [1967]. "Structure, Sign, and Play in the Discourse of the Human Sciences." In Jacques Derrida, *Writing and Difference,* translated by A. Bass, 278–93. London: Routledge.

Doane, Mary Ann. 1990. "Technophilia: Technology, Representation and the Feminine." In Mary Jacobus, Evelyn Fox Keller, and Sally Shuttleworth, eds., *Body/Politics: Women and the Discourse of Science,* 163–76. London: Routledge.

———. 1991a [1982]. "Film and the Masquerade: Theorizing the Female Spectator." In *Femmes Fatales: Feminism, Film Theory, Psychoanalysis,* 17–32. New York: Routledge.

———. 1991b [1988–89]. "Masquerade Reconsidered: Further Thoughts on the Female Spectator." In *Femmes Fatales: Feminism, Film Theory, Psychoanalysis,* 33–43. New York: Routledge.

———. 1991c. *Femmes Fatales: Feminism, Film Theory, Psychoanalysis.* New York: Routledge.

Dolar, Mladen. 1991. "I Shall Be with You on Your Wedding Night: Lacan and the Uncanny." *October* 58, fall: 5–24.

Donald, James. 1995. "The City, the Cinema: Modern Spaces." In C. Jenks, ed., *Visual Culture*, 77–95. London: Routledge.

Douglas, Mary. 1966. *Purity and Danger: An Analysis of Concepts of Pollution and Taboo*. New York: Praeger.

Duden, Barbara. 1993. *Disembodying Women: Perspectives on Pregnancy and the Unborn*. Translated by L. Hoinacki. Cambridge: Harvard University Press.

Dyer, Richard. 1982. "Don't Look Now: The Male Pin-Up." *Screen* 23, no. 3–4: 61–73.

———. 1993. "Coming Out as Going In: The Image of the Homosexual as a Sad Young Man." In Richard Dyer, *The Matter of Images: Essays on Representations*, 73–92. London: Routledge.

———. 1998. *Stars*. London: British Film Institute.

———. 2004. *Heavenly Bodies: Film Stars and Society*. London: Routledge.

———. 2007. *Pastiche: Knowing Imitation*. London: Routledge.

Eco, Umberto. 1987. *Travels in Hyperreality*. London: Picador.

Edelman, Lee. 2004. *No Future: Queer Theory and the Death Drive*. Durham, N.C.: Duke University Press.

Egorova, Yulia, Andrew Edgar, and Stephen Pattison. 2006. "The Meanings of Genetics: Accounts of Biotechnology in the Work of Habermas, Baudrillard and Derrida." *International Journal of the Humanities* 3, no. 2: 97–103.

Fletcher, John. 1988. "Versions of Masquerade." *Screen* 29, no. 3: 43–69.

Fortier, Anne-Marie. 2008. *Multicultural Horizons: Diversity and the Limits of the Civil Nation*. London: Routledge.

Foster, Gwendolyn Audrey. 2003. *Performing Whiteness: Postmodern Re/constructions in the Cinema*. Albany: State University of New York Press.

Foucault, Michel. 1973. *The Birth of the Clinic*. New York: Routledge.

———. 1978. *The History of Sexuality*. Vol. 1, *An Introduction*. Translated by R. Hurley. New York: Vintage.

Franklin, Sarah. 2000. "Life Itself: Global Nature and the Genetic Imaginary." In Sarah Franklin, Celia Lury, and Jackie Stacey, *Global Nature, Global Culture*, 188–227. London: Sage.

———. 2002. "Flat Life: Conception after Dolly." Paper presented at the Biotechnology, Philosophy and Sex Conference, Ljubljana, Slovenia, October 12.

———. 2007. *Dolly Mixtures: The Remaking of Genealogy*. Durham, N.C.: Duke University Press.

Franklin, Sarah, Celia Lury, and Jackie Stacey. 2000. *Global Nature, Global Culture*. London: Sage.

Franklin, Sarah, and Susan McKinnon. 2001. *Relative Values: Reconfiguring Kinship Studies*. Durham, N.C.: Duke University Press.

Franklin, Sarah, and Helena Ragoné. 1997. *Reproducing Reproduction: Kinship, Power, and Technological Innovation*. Philadelphia: University of Pennsylvania Press.

Freud, Sigmund. 1991 [1914]. "On Narcissism: An Introduction." In *On Metapsy-chology*, edited by Angela Richards, vol. 11 of *Penguin Freud Library*, 59–98. New York: Penguin.

———. 2002. [1918]. *The Wolfman and Other Cases*. Translated by L. A. Huish. London: Penguin Classics.

Friedberg, Anne. 1993. *Window Shopping: Cinema and the Postmodern*. Berkeley: University of California Press.

Fuss, Diana. 1994. "Interior Colonies: Frantz Fanon and the Politics of Identification." *Diacritics* 24, no. 2–3: 19–42.

George, Susan. 2001. "Not Exactly 'Of Woman Born': Procreation and Creation in Recent Science Fiction Films." *Journal of Popular Film and Television* 29, no. 4: 177–83.

Gibson, Pamela Church. 2001. "You've Been in my Life So Long I Can't Remember Anything Else": Into the Labyrinth with Ripley and the Alien." In Matthew Tinkcom and Amy Villarejo, eds., *Keyframes: Popular Cinema and Cultural Studies*, 35–51. London: Routledge.

Giddens, Anthony. 1990. *The Consequences of Modernity*. Cambridge, U.K.: Polity.

Gilloch, Graeme. 1996. *Myth and Metropolis: Walter Benjamin and the City*. Cambridge, Mass.: Polity.

Gledhill, Christine. 1991. *Stardom: Industry of Desire*. London: Routledge.

Gonder, Patrick. 2003. "Like a Monstrous Jigsaw Puzzle: Genetics and Race in Horror Films of the 1950s." *The Velvet Light Trap* 52: fall, 33–44.

Grosz, Elizabeth. 1991. "Freaks." *Social Semiotics* 1, no. 2: 22–38.

———. 1994. *Volatile Bodies: Towards a Corporeal Feminism*. Bloomington: Indiana University Press.

———. 1995. *Space, Time, and Perversion: Essays on the Politics of Bodies*. New York: Routledge.

Grusin, Richard A., and Jay David Bolter. 1999. *Remediation: Understanding New Media*. Cambridge: MIT Press.

Gu, Felicity Rose, and Zilai Tang. 2002. "Shanghai: Reconnecting to the Global Economy." In Saskia Sassen, ed., *Global Networks, Linked Cities*, 273–308. New York: Routledge.

Gumbrecht, Hans Ulrich, and Michael Marrinan, eds. 2003. *Mapping Benjamin: The Work of Art in the Digital Age*. Stanford, Calif.: Stanford University Press.

Gunew, Sneja. 2004. *Haunted Nations: The Colonial Dimensions of Multi-culturalisms*. London: Routledge.

Halberstam, Judith. 1998. *Female Masculinity*. Durham, N.C.: Duke University Press.

Hall, Stuart, Dorothy Hobson, Andrew Lowe, and Paul Willis, eds. 1980. *Culture, Media, Language: Working Papers in Cultural Studies, 1972–79*. London: Hutchinson.

Haran, Joan, et al. 2007. *Human Cloning in the Media*. London: Routledge.

Haraway, Donna. 1989. *Primate Visions: Gender, Race and Nature in the World of Modern Science*. New York: Routledge.

———. 1991. *Simians, Cyborgs, and Women: The Reinvention of Nature*. London: Free Association Books.

———. 1997. *Modest_Witness@Second_Millennium. FemaleMan_Meets_Onco Mouse*. New York: Routledge.

Hart, Lynda. 1994. *Fatal Women: Lesbian Sexuality and the Mark of Aggression*. Princeton, N.J.: Princeton University Press.

Harvey, Sylvia. 1998. "Woman's Place: The Absent Family in Film Noir." In E. Ann Kaplan (ed.), *Women in Film Noir*, rev. ed., 35–46. London: British Film Institute.

Hayles, N. Katherine. 1999. *How We Became Posthuman: Virtual Bodies in Cybernetics, Literature and Informatics*. Chicago: University of Chicago Press.

———. 2005. *My Mother Was a Computer: Digital Subjects and Literary Texts*. Chicago: University of Chicago Press.

Hemmings, Clare. 2005. "Invoking Affect: Cultural Theory and the Ontological Turn." *Cultural Studies* 19, no. 5: 548–67.

Hennion, Antoine, and Bruno Latour. 2003. "How to Make Mistakes on So Many Things at Once—and Become Famous for It." In Hans Ulrich Gumbrecht and Michael Marrinan, eds., *Mapping Benjamin: The Work of Art in the Digital Age*, 91–97. Stanford, Calif.: Stanford University Press.

Hills, Elizabeth. 1999. "From 'Figurative Males' to Action Heroines: Further Thoughts on Active Women in the Cinema." *Screen* 40, no. 1: 38–51.

Hills, Matthew. 2005. "The Generic Engineering of Monstrosity: Appropriations of Genetics in 1990s 'Species-Level Biohorror.'" Paper presented at the April "Cinema and Technology" Conference, Lancaster University.

Hoberman, J. 2004. "Reproductive Thriller Scores Genetic Coup, Flunks Chemistry." *Village Voice*, July 27. http:www.villagevoice.com/2004-07-27/film/.

Holloway, Wendy, Couze Venn, Valerie Wallerdine, and Julian Henriques. 1998. *Changing the Subject: Psychology, Social Regulation and Subjectivity*. London: Routledge.

Holmlund, Chris. 1993. "Masculinity as Multiple Masquerade: The 'Mature' Stallone and the Stallone Clone." In Steve Cohan and Ina Rae Hark, eds., *Screening the Male: Exploring Masculinity in Hollywood Cinema*, 213–29. London: Routledge.

———. 2002. *Impossible Bodies: Femininity and Masculinity at the Movies*. London: Routledge.

Hurley, Kelly. 1995. "Reading Like an Alien: Posthuman Identity in Ridley Scott's *Alien* and David Cronenberg's *Rabid*." In Judith Halberstam and Ira Livingston, eds., *Posthuman Bodies*, 203–24. Indianapolis: Indiana University Press.

Irigaray, Luce. 1985. *Speculum of the Other Woman*. Translated by G. C. Gill. Ithaca, N.Y.: Cornell University Press.

James, Joy. 2003. "Becoming Photographs: Aesthetics of Immanence." Ph.D. diss., University of British Columbia, Vancouver.

Jameson, Fredric. 1983. "Postmodernism and Consumer Society." In H. Foster, ed., *The Anti-Aesthetic: Essays on Postmodern Culture*, 111–26. Port Townsend, Wash.: Bay Press.

Jennings, Ros. 1995. "Desire and Design: Ripley Undressed." In T. Wilton, ed., *Immortal Invisible: Lesbians and the Moving Image*, 193–206. London: Routledge.

Johnston, Claire. 1975. "Femininity and the Masquerade: Anne of the Indies." In Claire Johnston and Paul Willemen, eds., *Jacques Tourneur*, 36–44. Edinburgh: Edinburgh Film Festival.

Jordanova, Ludmilla, and Peter Hulme. 1990. *The Enlightenment and Its Shadows*. London: Routledge.

Kaplan, E. Ann, ed. 1980. *Women in Film Noir*. London: British Film Institute.

———, ed. 1998. *Women in Film Noir*. Rev. ed. London: British Film Institute.

Kavanagh, James. 1990. "Feminism, Humanism and Science in Alien." In Annette Kuhn (ed.), 73–81. *Alien Zone: Cultural Theory and Contemporary Science Fiction Cinema*. London: Verso.

Keller, Evelyn Fox. 1995. *Refiguring Life: Metaphors of Twentieth-Century Biology*. New York: Columbia University Press.

———. 2000. *The Century of the Gene*. Cambridge: Harvard University Press.

Kember, Sarah. 1996. "Feminism, Technology and Representation." In James Curran, David Morley, and Valerie Walkerdine, eds., *Cultural Studies and Communication*, 229–47. London: Edward Arnold.

———. 1998. *Virtual Anxiety: Photography, New Technologies and Subjectivity*. Manchester: Manchester University Press.

———. 2003. *Cyberfeminism and Artificial Life*. London: Routledge.

Khanna, Ranjana. 2003. *Dark Continents: Psychoanalysis and Colonialism*. Durham, N.C.: Duke University Press.

Kinder, Marsha. 2005. "A Cinema of Intelligent Agents." In Meredith Tromble, ed., *Art and Films of Lynn Hershman-Leeson: Secret Agents, Private I*, 169–82. Berkeley: University of California Press.

Kirby, David. 2000. "The New Eugenics in Cinema: Genetic Determinism and Gene Therapy in GATTACA." *Science Fiction Studies* 27: 193–215.

Kirkham, Pat, and Janet Thumim. 1993. *You Tarzan: Masculinity, Movies and Men*. London: Lawrence and Wishart.

Kristeva, Julia. 1982. *Powers of Horror: An Essay on Abjection*. Translated by L. S. Roudiez. New York: Columbia University Press.

Kuhn, Annette, ed. 1990. *Alien Zone: Cultural Theory and Contemporary Science Fiction Cinema*. London: Verso.

———. 1994. *Women's Pictures: Feminism and Cinema*. London: Verso.

———, ed. 1999. *Alien Zone II: The Spaces of Science Fiction Cinema*. London: Verso.

Lai, Larissa. 2004. "Sites of Articulation: Robyn Morris in Conversation with Larissa Lai." *West Coast Line No. 44* 38, no. 2: 20–30.

Landecker, Hannah. 2005. "Cellular Features: Microcinematography and Film Theory." *Critical Inquiry* 31: 903–37.

———. 2007. *Culturing Life: How Cells Became Technologies*. Cambridge: Harvard University Press.

Latour, Bruno, and Peter Weibel. 2005. *Iconoclash: Beyond the Image Wars in Science, Religion and Art*. Cambridge: MIT Press.

Lauritsen, John. 1993. "Political-economic Construction of Gay Male Cloning Identity." *Journal of Homosexuality* 24, nos. 3-4: 221–32.

Le Breton, David. 2004. "Genetic Fundamentalism or the Cult of the Gene." *Body and Society* 10, no. 4: 1–20.

Lehman, Peter, ed. 2001. *Masculinity: Bodies, Movies, Culture*. New York: Routledge.

Levin, Maud. 1993. *Cut with a Kitchen Knife: The Weimar Photomontages of Hannah Höch*. New Haven, Conn.: Yale University Press.

Lévi-Strauss, Claude. 1969. *The Elementary Structures of Kinship*. Rev. ed. Edited by Rodney Needham, translated by James Harle Bell. Boston: Beacon.

Lippit, Akira M. 2005. *Atomic Light (Shadow Optics)*. Minneapolis: University of Minnesota Press.

Low, Gail Ching-Liang. 1996. *White Skins/Black Masks: Representation and Colonialism*. London: Routledge.

Lundin, Michael. 2002. "(Wo)man-machine-insect: Posthuman Bodies in Jean-Pierre Jeunet's *Alien: Resurrection* and Philip K. Dick's *Ubik*." Unpublished paper.

———. n.d. "Spiders from Mars: Refiguring Subjectivity in the Novels of Philip K. Dick, 1953–66." Ph.D. diss., Stockholm University.

Luyken, Gunda. 2004. "Hannah Höch's Album: Collection of Materials, Sketchbook, or Conceptual Art?" In Hatje Cantz, ed., *Hannah Höch's Album*. Ostfildern-Ruit, Germany: Hatje Cantz Verlag.

Mackenzie, Adrian. 2002. *Transductions: Bodies and Machines at Speed*. New York: Continuum.

Manovich, Lev. 1999. "What Is Digital Cinema?" In Peter Lunenfield, ed., *The Digital Dialectic: New Essays on New Media*, 172–92. Cambridge: MIT Press.

———. 2001. *The Language of New Media*. Cambridge: MIT Press.

Marchessault, Janine, and Kim Sawchuk, eds. 2000. *Wild Science: Reading Feminism, Medicine and the Media*. New York: Routledge.

Marks, Laura. 2000. *The Skin of the Film: Intercultural Cinema, Embodiment, and the Senses*. Durham, N.C.: Duke University Press.

McNeil, Maureen. 1987. *Under the Banner of Science: Erasmus Darwin and His Age*. Manchester: Manchester University Press.

———. 2000. "Techno-triumphalism, Techno-tourism, American Dreams and Feminism." In Sara Ahmed et al., eds., *Transformations: Thinking Through Feminism*, 221–34. New York and London: Routledge.

Merck, Mandy. 2000. "The Lesbian Hand." In *In Your Face: 9 Sexual Studies*, 124–27. New York: New York University Press.

Mitchell, W. J. T. 1992. *The Reconfigured Eye*. Cambridge: MIT Press.

———. 2005. *What Do Pictures Want? The Lives and Loves of Images*. Chicago: Chicago University Press.

Modleski, Tania. 1988. *The Women Who Knew Too Much: Hitchcock and Feminist Theory*. New York: Methuen.

Morris, Meagan. 1992. "Great Moments in Social Change: King Kong and the Human Fly." In Beatriz Colomina, ed., *Sexuality and Space*, 1–52. New York: Princeton Architectural Press.

Moseley, Rachel. 2005. *Fashioning Film Stars: Dress, Culture, Identity*. London: British Film Institute.

Mulvey, Laura. 1989. *Visual and Other Pleasure*. London: Macmillan.

———. 1996. *Fetishism and Curiosity*. Bloomington: Indiana University Press.

———. 2006. *Death 24x a Second: Stillness and the Moving Image*. London: Reaktion Books.

Neale, Steve. 1983. "Masculinity as Spectacle." *Screen* 24, no. 6: 2–16.

Nelkin, Dorothy, and M. Susan Lindee. 1995. *The DNA Mystique: The Gene as a Cultural Icon*. New York: W. H. Freeman.

Newman, Kim, ed. 2002. *Science Fiction/Horror: A Sight and Sound Reader*. London: British Film Institute.

Newton, Esther. 1972. *Mother Camp: Female Impersonators in America*. Chicago: University of Chicago Press.

Newton, Judith. 1990. "Feminism and Anxiety in *Alien*." In Annette Kuhn, ed., *Alien Zone: Cultural Theory and Contemporary Science Fiction Cinema*, 82–90. London: Verso.

Nottingham, Stephen. 2000. *Screening DNA: Exploring the Cinema-Genetics Interface*. (Book on CD, DNA Books.)

Orbaugh, Sharalyn. 2005. "The Genealogy of the Cyborg in Japanese Popular Culture." In Wong Kin Yuen, Gary Westfahl, and Amy Kit-sze Chan, eds., *World Weavers: Globalization, Science Fiction, and Cybernetic Revolution*, 55–72. Hong Kong: Hong Kong University Press.

O'Riordan, Kate. 2006. "Women in Media Debates about Cloning: Cinematic Representations." Paper presented in the Department of Media and Film, University of Sussex, May 17.

Ostherr, Kirsten. 2005. *Cinematic Prophylaxis: Globalization and Contagion in the Discourse of World Health*. Durham, N.C.: Duke University Press.

Pearce, Lynne. 1994. *Reading Dialogics*. London: Edward Arnold.

———. 2003. *The Rhetorics of Feminism: Readings in Contemporary Literary Theory and the Popular Press*. London: Routledge.

———. 2007. *Romance Writing*. Cambridge, U.K.: Polity.

Penley, Constance. 1991. "Time Travel, Primal Scene and the Critical Dystopia." In Constance Penley et al. (eds.), *Close Encounters: Film, Feminism, and Science Fiction*, 63–82. Minneapolis: University of Minnesota Press.

Penley, Constance, Elisabeth Lyon, Lynn Spigel, and Janet Bergstrom, eds. 1991. *Close Encounters: Film, Feminism, and Science Fiction*. Minneapolis: University of Minnesota Press.

Phillips, Adam. 1998. "Sameness Is All." In Martha C. Nussbaum and Cass R. Sunstein, eds., *Clones and Clones: Facts and Fantasies about Human Cloning*, 88–94. New York: W. W. Norton.

Rabinow, Paul. 1999. *French DNA: Trouble in Purgatory*. Chicago: University of Chicago Press.

Reilly, Philip R. 2004. "Divining DNA." In Suzanne Anker and Dorothy Nelkin, eds., *The Molecular Gaze: Art in the Genetic Age*, xi–xvi. Woodbury, N.Y.: Cold Spring Harbor Laboratory Press.

Rich, Adrienne. 1980. "Compulsory Heterosexuality and Lesbian Existence." *Signs* 5, no. 4 (summer): 631–60.

Ritter, Henning. 2003. "Toward the *Artwork* Essay, Second Version." In Hans Ulrich Gumbrecht and Michael Marrinan, eds., *Mapping Benjamin: The Work of Art in the Digital Age*, 203–10. Stanford, Calif.: Stanford University Press.

Riviere, Joan. 1986 [1929]. "Womanliness as a Masquerade." In Victor Burgin, James Donald, and Cora Kaplan, eds., *Formations of Fantasy*, 35–44. New York: Methuen.

Roof, Judith. 1996a. *Come as You Are: Sexuality and Narrative*. New York: Columbia University Press.

———. 1996b. *Reproductions of Reproduction: Imaging Symbolic Change*. New York: Routledge.

———. 2007. *The Poetics of DNA*. Minneapolis: University of Minnesota Press.

Rose, Nikolas. 1996. *Inventing Our Selves: Psychology, Power, and Personhood*. Cambridge: Cambridge University Press.

Rothman, Barbara Katz. 1998. *Genetic Maps and Human Imaginations: The Limits of Science in Understanding Who We Are*. New York: W. W. Norton.

Said, Edward. 1979. *Orientalism*. New York: Vintage.

Sassen, Saskia, ed. 2002. *Global Networks, Linked Cities*. New York: Routledge.

Sawchuk, Kim. 2000. "Biotourism, Fantastic Voyage, and Sublime Inner Space." In Janine Marchessault and Kim Sawchuk, eds., *Wild Science: Reading Feminism, Medicine, and the Media*, 9–23. New York: Routledge.

Schmidt, Siegfried J. 2003. "From Aura-Loss to Cyberspace: Further Thoughts on Walter Benjamin." In Hans Ulrich Gumbrecht and Michael Marrinan (eds.), *Mapping Benjamin: The Work of Art in the Digital Age*, 79–82. Stanford, Calif.: Stanford University Press.

Schor, Naomi. 1995. *Bad Objects: Essays Popular and Unpopular.* Durham, N.C.: Duke University Press.

Schwartz, Hillel. 1996. *The Culture of the Copy: Striking Likenesses, Unreasonable Facsimiles.* New York: Zone.

Sedgwick, Eve Kosofsky. 1985. *Between Men: English Literature and Male Homosocial Desire.* New York: Columbia University Press.

——. 1994. *Epistemology of the Closet.* London: Penguin.

Shiff, Richard. 2003. "Digitized Analogies." In Hans Ulrich Gumbrecht and Michael Marrinan, eds., *Mapping Benjamin: The Work of Art in the Digital Age,* 63–70. Stanford, Calif.: Stanford University Press.

Shingler, Martin. 1995. "Masquerade or Drag? Bette Davis and the Ambiguities of Gender." *Screen* 36, no. 3: 179–92.

Shohat, Ella. 1992a. "'Lasers for Ladies': Endodiscourse and the Inscription of Science." *Camera Obscura* 29: 57–90.

——. 1992b. "Notes on the 'Post Colonial.'" *Social Text* 31/32: 99–113.

——, ed. 1998. *Talking Visions: Multicultural Feminism in a Transnational Age.* New York: New Museum of Contemporary Art and Cambridge: MIT Press.

Silverman, Kaja. 1991. "Back to the Future." *Camera Obscura* 27: 115.

——. 1996. *The Threshold of the Visible World.* New York: Routledge.

Smelik, Anneke. 2008. "Tunnel Vision: Inner, Outer and Virtual Space in Science Fiction Film and Medical Documentaries." In Anneke Smelik and Nina Lykke, eds., *Bits of Life: Feminism at the Intersections of Media, Bioscience and Technology,* 129–46. Seattle: University of Washington Press.

Smelik, Anneke, and Nina Lykke, eds. 2008. *Bits of Life: Feminism at the Intersections of Media, Bioscience and Technology.* Seattle: University of Washington Press.

Sobchack, Vivian C. 1980. *The Limits of Infinity: The American Science Fiction Film, 1950–75.* New York: A. S. Barnes.

——. 1997. *Screening Space: The American Science Fiction Film.* New Brunswick, N.J.: Rutgers University Press.

Sontag, Susan. 1977. *Illness as Metaphor.* London: Penguin.

Spielmann, Yvonne. 1999. "Aesthetic Features in Digital Imaging: Collage and Morph." *Wide Angle* 21, no. 1: 131–48.

Springer, Claudia. 1996. *Electronic Eros: Bodies and Desire in the Postindustrial Age.* London: Athlone.

Squier, Susan. 1999. "Negotiating Boundaries: From Assisted Reproduction to Assisted Replication." In E. Ann Kaplan and Susan Squier, eds., *Playing Dolly: Technocultural Formations, Fantasies and Fictions of Assisted Reproduction,* 101–15. New Brunswick, N.J.: Rutgers University Press.

Stacey, Jackie. 1997. *Teratologies: A Cultural Study of Cancer.* London: Routledge.

Stafford, Barbara Maria. 1991. *Body Criticism: Imaging the Unseen in Enlightenment Art and Medicine*. Cambridge: MIT Press.

Steinberg, Deborah Lynn. 1997. *Bodies in Glass: Genetics, Eugenics, Embryo Ethics*. Manchester: Manchester University Press.

Stewart, Susan. 1993. *On Longing: Narratives of the Miniature, the Gigantic, the Souvenir, the Collection*. Durham, N.C.: Duke University Press.

Straayer, Chris. 2001. "The Talented Poststructuralist: Heteromasculinity, Gay Artifice and Class Passing." In Peter Lehman, ed., *Masculinity: Bodies, Movies, Culture*, 115–32. New York: Routledge.

Strathausen, Carsten. 2003. "Uncanny Spaces: The City in Ruttman and Vertov." In M. Shiel and T. Fitzmaurice (eds.), *Screening the City*, 15–40. New York: Verso.

Studlar, Gaylyn. 1996. *This Mad Masquerade: Stardom and Masculinity in the Jazz Age*. New York: Columbia University Press.

Swift, Jonathan. 2001. *Gulliver's Travels*. London: Penguin.

Szope, Dominika. 2003. "Peter Weibel." In Jeffrey Shaw and Peter Weibel, eds., *Future Cinema: The Cinematic Imaginary After Film*, 180–192. Cambridge: MIT Press.

Tasker, Yvonne. 1993. *Spectacular Bodies: Gender, Genre and the Action Cinema*. London: Routledge.

———. 1998. *Working Girls: Gender and Sexuality in Popular Cinema*. London: Routledge.

Telotte, J. P. 1990. "The Double of Fantasy and the Space of Desire." In Annette Kuhn, ed., *Alien Zone: Cultural Theory and Contemporary Science Fiction Cinema*, 152–59. London: Verso.

———. 1995. *Replications: A Robotic History of the Science Fiction Film*. Urbana: University of Illinois Press.

———. 1999. *A Distant Technology: Science Fiction Film and the Machine Age*. Hanover, N.H.: Wesleyan University Press.

———. 2001. *Science Fiction Film*. Cambridge: Cambridge University Press.

Thacker, Eugene. 2004. *Biomedia*. Minneapolis: University of Minnesota Press.

Thompson, Michael. 1995. "The Representation of Life." In Rosalind Hursthouse, Gavin Lawrence, and Warren Quinn, eds., *Virtues and Reasons: Philippa Foot and Moral Theory*, 248–96. Oxford: Clarendon Press of Oxford University Press.

Thornham, Sue. 1997. *Passionate Detachments: An Introduction to Feminist Film Theory*. London: Edward Arnold.

———, ed. 1999. *Feminist Film Theory: A Reader*. Edinburgh: Edinburgh University Press.

Treichler, Paula A., Lisa Cartwright, and Constance Penley, eds. 1998. *The Visible Woman: Imaging Technologies, Gender, and Science*. New York: New York University Press.

Tromble, Meredith, ed. 2005. *The Art and Films of Lynn Hershman-Leeson: Secret Agents, Private I.* Berkeley: University of California Press.

Turney, Jon. 1998. *Frankenstein's Footsteps: Science, Genetics and Popular Culture.* New Haven, Conn.: Yale University Press.

Tyler, Carole-Ann. 2003. *Female Impersonation.* New York: Routledge.

Tyler, Imogen. 2001. "Skin-tight: Celebrity, Pregnancy and Subjectivity." In Sara Ahmed and Jackie Stacey, eds., *Thinking Through the Skin*, 69–84. London: Routledge.

van Dijck, José. 1998. *Imagenation: Popular Images of Genetics.* New York: New York University Press.

———. 2005. *The Transparent Body: A Cultural Analysis of Medical Imaging.* Seattle: University of Washington Press.

Varmus, Harold, and Robert A. Weinberg. 1993. *Genes and the Biology of Cancer.* New York: Scientific American Library.

Vera, Hernán, and Andrew M. Gordon. 2003. *Screen Saviors: Hollywood Fictions of Whiteness.* Lanham, Md.: Rowman and Littlefield.

Vidler, Anthony. 1992. *The Architectural Uncanny: Essays in the Modern Unhomely.* Cambridge: MIT Press.

———. 2000. *Warped Space: Art, Architecture, and Anxiety in Modern Culture.* Cambridge: MIT Press.

Waldby, Catherine. 2000. *The Visible Human Project: Informatic Bodies and Posthuman Medicine.* London: Routledge.

Watson, James D. 1968. *The Double Helix: A Personal Account of the Discovery of the Structure of DNA.* New York: Atheneum.

Weaver, Sigourney. 1997. Interview in *Film Review*, December, 34–39.

Weber, Cynthia. 2006. *Imagining America at War: Morality, Politics and Film.* London: Routledge.

Weibel, Peter. 1996. "Postontologische Kunst, 1994." In Romana Schuler and Peter Weibel, eds., *Peter Weibel: Bildwelten 1982–1996.* Vienna: Triton.

Weimar, Klaus. 2003. "Textual-Critical Remarks et Alia." In Hans Ulrich Gumbrecht and Michael Marrinan (eds.), *Mapping Benjamin: The Work of Art in the Digital Age*, 188–94. Stanford, Calif.: Stanford University Press.

Weinbaum, Alys Eve. 2008. "Racial Aura: Walter Benjamin and the Work of Art in a Biotechnological Age." *Literature and Medicine* 26, no. 1: 207–39.

Weston, Kath. 1991. *Families We Choose: Lesbians, Gays, Kinship.* New York: Columbia University Press.

———. 1998. *Long Slow Burn: Sexuality and Social Science.* New York: Columbia University Press.

Wiegman, Robyn. 2006. "Heteronormativity and the Desire for Gender." *Feminist Theory* 7, no. 1: 89–104.

Willemen, Paul. 2002. "Reflections on Digital Imagery: Of Mice and Men." In M. Rieser and A. Zapp, eds., *New Screen Media: Cinema/Art/Narrative*, 14–29. London: British Film Institute.

Williams, Raymond. 1975. *The Country and the City*. St. Albans, Hertfordshire, U.K.: Paladin.

Wolfram, Eddie. 1975. *History of Collage: An Anthology of Collage, Assemblage and Event Structures*. London: Studio Vista.

Wollen, Peter. 2002. *Paris Hollywood: Writings on Film*. New York: Verso.

Wood, Aylish. 2002. *Technoscience in Contemporary American Film: Beyond Science Fiction*. Manchester: Manchester University Press.

Young, Lola. 1996. "'Miscegenation' and the Perils of 'Passing': Films from the 1950s and 1960s." In *Fear of the Dark: "Race," Gender, and Sexuality in the Cinema*, 84–114. London: Routledge.

Filmography

Alien. 1979. Directed by Ridley Scott.

Aliens. 1986. Directed by James Cameron.

Alien 3. 1992. Directed by David Fincher.

Alien: Resurrection. 1997. Directed by Jean-Pierre Jeunet.

The Andromeda Strain. 1971. Directed by Robert Wise.

Artificial Intelligence. 2001. Directed by Steven Spielberg.

Blade Runner. 1982. Directed by Ridley Scott.

Brief Encounter. 1945. Directed by David Lean.

Casshern. 2004. Directed by Kazuaki Kiriya.

Chinatown. 1974. Directed by Roman Polanski.

Code 46. 2003. Directed by Michael Winterbottom.

Come Back to the Five and Dime Jimmy Dean, Jimmy Dean. 1982. Directed by Robert Altman.

Conceiving Ada. 1997. Directed by Lynn Hershman-Leeson.

Contact. 1997. Directed by Roger Zemeckis.

Deep Blue Sea. 1999. Directed by Renny Harlin.

DNA. 1997. Directed by William Mesa.

Double Indemnity. 1944. Directed by Billy Wilder.

The Eighteenth Angel. 1998. Directed by William Bindley.

Equilibrium. 2002. Directed by Kurt Wimmer.

Eternal Sunshine of the Spotless Mind. 2004. Directed by Michel Gondry.

Evolution. 2001. Directed by Ivan Reitman.

eXistenZ. 1999. Directed by David Cronenberg.

The Fifth Element. 1997. Directed by Luc Besson.

The Fly. 1958. Directed by Kurt Neumann.

The Fly. 1986. Directed by David Cronenberg.

Frankenstein. 1931. Directed by James Whale.

Gattaca. 1997. Directed by Andrew Niccol.

Genetic Admiration. 2005. Directed by Frances Leeming.

Godsend. 2004. Directed by Nick Hamm.

Honey, I Shrunk the Kids. 1989. Directed by Joe Johnston.

Hulk. 2003. Directed by Ang Lee.

The Human Body. 1998. TV series. Directed by Christopher Spencer.

In This World. 2002. Directed by Michael Winterbottom.

The Incredible Shrinking Man. 1957. Directed by Jack Arnold.

Invasion of the Body Snatchers. 1956. Directed by Don Siegel.

The Island. 2005. Directed by Michael Bay.

The Island of Dr. Moreau. 1996. Directed by John Frankenheimer.

Jurassic Park. 1993. Directed by Steven Spielberg.

The Lady from Shanghai. 1947. Directed by Orson Welles.

The Last Time I Saw Paris. 1954. Directed by Richard Brooks.

Lorenzo's Oil. 1992. Directed by George Miller.

The Man with the Golden Arm. 1955. Directed by Otto Preminger.

Memento. 2000. Directed by Christopher Nolan.

Mimic. 1997. Directed by Guillermo del Toro.

Mimic 2. 2001. Directed by Jean de Segonzac.

Minority Report. 2002. Directed by Steven Spielberg.

Multiplicity. 1996. Directed by Harold Ramis.

1984. 1956. Directed by Michael Anderson.

Now, Voyager. 1942. Directed by Irving Rapper.

Out of the Past. 1947. Directed by Jacques Tourneur.

The Relic. 1997. Directed by Peter Hymans.

Robocop. 1987. Directed by Paul Verhoeven.

Rope. 1948. Directed by Alfred Hitchcock.

The 6th Day. 2000. Directed by Roger Spottiswoode.

Species. 1995. Directed by Roger Donaldson.

Species II. 1998. Directed by Peter Medak.

Star Wars: Episode II—Attack of the Clones. 2002. Directed by George Lucas.

Stingray. 1964-65. TV series. Directed by Alan Pattillo et al.

Swoon. 1992. Directed by Tom Kalin.

The Talented Mr. Ripley. 1999. Directed by Anthony Minghella.

Teknolust. 2002. Directed by Lynn Hershman-Leeson.

Thriller. 1983. Directed by John Landis.

Thunderbirds. 1965. TV series. Directed by Desmond Saunders et al.

Toy Story. 1995. Directed by John Lasseter.

The Twilight of the Golds. 1997. Directed by Ross Kagan Marks.

Vertigo. 1958. Directed by Alfred Hitchcock.

Index

Abject, 37, 50–53, 61–62, 107, 158, 284
    n. 8; body, 49, 113, 261; desire, 21;
    genetic, 37–38, 48–49, 64; imagi-
    nary, xiii; psychoanalytic theory
    of, 19, 47–50, 52, 54–56. *See also*
    Creed, Barbara; Kristeva, Julia
Aesthetic, 6, 43, 151, 167, 171, 190, 215,
    219, 222, 231, 240–41, 245, 255–56;
    abstraction, 115; art-house, x;
    citational, 198, 219, 283 n. 14; delay
    and, 267; desire and, 223; digital, 19,
    31, 193; Easternized, 200; eugenic,
    124; generative, xii, 217; genetic, 6,
    153, 200, 217–18, 227; imaginary, 10;
    intertextual, 283 n. 14; mimetic, 255;
    performative, 240; repetition, 60,
    143; seductive, 113; technology and,
    212, 223; visual, 200
Affect, x, 4, 89, 100, 129, 161–62,
    182–83, 204–5, 212, 221–23, 270,

283 n. 17, 283 n. 18; authenticity of,
    15; bio-aura and, 182–83, 187–88;
    biomolecular, 283 n. 17; body, 222;
    clone and, 19, 21, 35, 162; genetic, 21,
    35, 146, 162, 169, 182, 186, 224; kin-
    ship, 44; power of, 10, 15; sameness
    and, 35; sexual deviance and, 28;
    transmission of, 21, 182–83. *See also*
    Benjamin, Walter; Brennan, Teresa
Agency, 22, 56, 71, 104, 122, 136,
    138–39, 145, 160, 174, 176, 207, 227,
    230, 248, 250–53; autonomous, 123;
    double, 127; genetic, 251; human, 71,
    228; individual, 91, 127; masculine,
    129, 236; in narrative, 251
Ahmed, Sara, 142, 279 n. 23, 283 n. 18
A.I. Artificial Intelligence, 140, 279
    n. 21
Alice in Wonderland, 242, 284 n. 7
Alien, 38, 52, 57, 277

Alien 3, 38, 49

Alien: Resurrection, x, 13–15, 36–65, 72, 74, 77, 82, 117, 150, 276 n. 21, 277 n. 10, 277 n. 11

Aliens, 38, 48, 50–53

Altman, Robert, 212

Analogy, 3, 27–8, 57, 65, 100, 147, 179–80, 217, 225–26, 240, 251, 255–56, 281 n. 2

Androgyny, 38, 56–57, 114, 119, 253, 276 n. 18

Andromeda Strain, The, 77

Angerer, Marie-Louise, 126, 223, 283 n. 18

Animal, 27, 39, 48, 51, 97, 154, 226, 255–56

Animate vs. inanimate, 196, 203, 229, 252, 260, 268–70

Animation, 189, 227–31, 234, 240, 250, 252–3, 256, 259, 270; artist, 227; biomediated, 220; cellular, 16; in film, 227–28, 236; genealogy of, 227; human, 260; mechanical, 260; potential, 197. *See also* Collage animation; *Genetic Admiration*

Anker, Suzanne, 4–5

Anxiety, 9, 33, 47, 72, 92, 100, 119, 131, 149, 151, 156, 168, 186, 268, 279 n. 20

Appadurai, Arjun, 142

Architecture, xi, 6, 3–4, 78–80, 137, 138, 141, 144, 147, 156, 175, 228, 242, 255; design, 243; monumental, 7, 123; postmodern, 139; security and, 140; urban, 139–40

Arnold, Jack, 284 n. 7

Artifice, x-xiii, 7, 22, 75, 77, 82, 102, 108, 114, 116, 128, 136, 152, 189, 196, 203, 218, 229, 256; desire and, xi, 13; in digital cultures, 8; femininity and, 13, 85, 104, 119–20, 198, 200, 214; genetic, 14, 212, 221, 227, 265; masculine, 118–20; sexual, 212

Artificial insemination, 31–32

Attachment, 102, 105, 174, 182, 199, 219, 229, 260; masculine, 105; narcissistic, 101; Oedipal, 172; unconscious, 11, 57

Audience, 3, 89, 102, 104, 110, 113, 115–20, 135, 189–91, 197, 233, 248, 253, 259, 261, 266; cinema, 81, 83, 116, 198, 207, 216; diegetic, 78, 81; expectations of, 52

Augé, Marc, 142–43

Aura, xii-xiii, 14–15, 182–87, 190, 194, 225, 240, 245, 264. *See also* Benjamin, Walter; Bio-aura

Austin, J. L., 256

Authenticity, 10, 15, 22–23, 110, 121, 128, 133, 170, 184, 186, 190, 193; of body, 89, 111, 182, 188, 194, 218, 223; desire for, 14; human and identity, 26, 72, 118; of image, 152, 180, 188, 194; of individual, 48, 65; lack of, 14; loss of, xii, 8, 183, 189, 193; masculine, 104, 129; threat to, 6, 162

Authorship, 71, 91, 126, 136, 194

Automaton, 20, 22, 215, 270; self-replicating, 198

Autonomy, 10, 53, 118, 123, 125–27, 253, 261, 266

Baecker, Dirk, 184, 194

Barthes, Roland, 99–100, 190, 206; *Camera Lucida*, 99; photography, 101, 189

Battaglia, Debbora, 36–38, 49, 275 n. 1, 276 n. 12; on clone figure as supplement, 14, 36, 275 n. 2

Battersby, Christine, 276 n. 15, 277 n. 1

Baudrillard, Jean, xii, 13, 195, 207, 215, 222, 233, 275 n. 6; on cancer, 27; on cloning, x, 19–34, 221, 234; on degree xerox of the species, 22; *Écriture automatique*, 22; on genetic engineering, 20–23, 17, 33; on genetic project; 26; on hell of the same, x, 23, 26; on kinship futures, 13; on paradigm of sameness, 21, 30, 35. *See also* Edelman, Lee

Bay, Michael, 118, 140

Bellour, Raymond, 180

Benjamin, Walter, 95, 166, 173; affect, 182; on aura, xii, 14, 181–88, 225; on cinema, 190, 269; on commodity fetish, 186, 244–46; on neurasthenia, 174–75; on photography, 183; on vagabondage, 174–75; "The Work of Art in the Age of Mechanical Reproduction," 180–88, 281 n. 1. *See also* Vidler, Anthony

Bergstrom, Janet, 276 n. 18

Berlant, Lauren, on the female complaint, 233

Bernardi, Daniel, 274 n. 22

Bersani, Leo, 11

Besson, Luc, 73, 140

Bhabha, Homi, on colonial mimicry, 106

Bindley, William, 39

Bio-aura, xiii, 15, 179, 181–83, 186–88, 193, 225. *See also* Benjamin, Walter

Biohorror, species level of, 39, 87

Biology, xi, 5–6, 13, 23, 25–26, 29–31, 33–34, 55, 69–71, 87, 160–61, 172, 186, 194, 225, 229, 259, 251; biological and cultural reproduction, 9, 22, 24, 63; biological code, xi, 85, 88, 90; detraditionalized, 188; deviant, 49; malformation and, 27; matter and, 15, 147; microbiology, 204, 282 n. 9; purity of, xi, 74; queering of, x, 19; recombinant forms in, xii; reproducibility in, 15, 28, 187, 192; unnatural, 56, 81. *See also* Cell: biology of; Cloning; Copy; Genetic engineering

Biomedia, 186, 217, 282 n. 12

Biomimicry, 95–112

Biotechnology, 15, 70, 202–3, 205

Birth, 48, 50–51, 63, 107, 192, 236, 241–42, 276 n. 15; of child, 187. *See also under* Monstrosity

Black, Karen, 199, 212

Blade Runner, 36, 42, 47, 140, 147, 159, 276 n. 9, 278 n. 4, 278 n. 9, 281 n. 18

Blood tie, 177, 130. *See also* Kinship

Body: clean and proper, 53; encoded, 72; envisioning of, 7, 264–65; geneticization and, ix, 6, 8, 68, 70, 73, 77, 80, 102, 109, 111, 122, 138–39, 147, 149, 151, 179–83, 198; grotesque, 43, 241, 256; human, 7–8, 38, 46–47, 64, 70, 79, 95, 111–12, 121, 137–38, 181, 232, 235–36; informal components of, 8, 63, 68; informationalization of, 11; interior, 5, 78, 138, 151, 236–37, 273 n. 4; mutability of, 16, racialized, xi, 73, 108, 160; recombinant, 284 n. 12; sexualized, xi; spectacle of, x, 53; transparent, 79, 137–38, 151. *See also* Female: body of; Genetic engineering; Kristeva, Julia; Monstrosity; van Dijck, José; White(s)

Body-horror film, 37, 54

Body rebellion film, 72

Bolter, David Jay, 152; on remediation, 282 n. 12

Boss, Pete, 47

Boundary, 39, 54, 97, 107, 158, 166, 200, 256, 270; collapse of, 61; of human body, 8, 64, 68, 99; of life and death, 7, 29, 49, 196, 261, 268–69; of monstrous and proper bodies, 38, 50; of sameness and difference, 36; of self and other, 102, 261; of white body, 75

Braidotti, Rosi, 40–41, 273 n. 4, 276 n. 8, 277 n. 9, 279 n. 17, 283 n. 18;

Braidotti, Rosi (*cont.*)
  teratological imagination, 37, 284
  n. 8
Brennan, Teresa, on transmission of
  affect, 21, 182–83, 283 n. 18. *See also*
  Affect
Brief Encounter, 147
Brooks, Richard, 206
Brophy, Philip, 47
Bruno, Giuliana, 140, 273 n. 4, 274
  n. 13, 282 n. 5
Brunsdon, Charlotte, 166, 277 n. 5,
  277 n. 6
Buchanan, Ian, 283 n. 18
Bukatman, Scott, 276 n. 9, 276 n. 14
Butler, Judith, 29–30, 54, 67, 118, 128,
  167–68, 194, 218, 240, 275 n. 9, 275
  n. 10, 275 n.11, 279 n. 18, 279 n. 19;
  on the abject, 158; on biological
  cloning 96; on difference, 26; on
  femininity, 86; on gender impersonation, 127; on heterosexual matrix,
  30, 32, 110–11, 188; on imitation,
  216; on masculine subject, 125–26;
  on masquerade, 12; on surface politics of the body, 55; on symbolic
  order, 33

Caillois, Roger, 99; on mimicry, 97
Cameron, James, 38
Camouflage, xi, 80, 84, 91, 97
Cancer, 29, 284 n. 2; as C word, 28;
  deviant cells in, 27; metastasis in,
  28, 221; mutation in, 83. *See also*
  Metastasis; Teratoma
Capitalism, 22, 43, 62, 146, 159
Carnival, 241, 244, 256. *See also*
  Stewart, Susan
Carroll, Lewis, 284 n. 7
Cartwright, Lisa, 7, 273 n. 4, 284 n. 6,
  284 n. 9
Casshern, 74, 140
Castañeda, Claudia, 4, 247–48

Cell: abnormal growth of, 28; behavior of, 41, 241; biology of, 229, 248;
  division of, 25, 27–28, 44; formation of, 26, 40; implantation of, 194;
  life, 46, 226; malfunction of, 45;
  manipulation of, 40, 84; mutation
  of, 39–42
Celluloid, 197, 208, 211, 214–16, 217,
  221, 224, 259, 267–68
Centre for Genetic Admiration, 228,
  242, 244, 246–47. See also *Genetic
  Admiration*
Child, 66–67, 95–97, 105, 242,
  246–49, 261, 275 n. 10. *See also
  under* Monstrosity; Mother
Childbirth. *See* Birth
Chinatown, 281 n. 17, 282 n. 13
Citation, xii, 53, 198, 217–8, 232, 255
Civilization, 30, 33, 73, 230, 252
Clairvoyance, 228; genetic, 162–64, 166
Class, 73–75, 116, 129, 131, 246
Cloning: advent of, 20–22, 25, 150,
  265, 274 n. 15, 274 n. 18, 274 n. 19,
  275 n. 2; digital, 198–200, 202, 210,
  218, 221–23; disease and, 27–28;
  failure of, 42, 45–46; figure of,
  24, 27, 36, 38, 54, 95–96, 104–11,
  263–64, 269; of gay male, 276 n. 19;
  gay subculture and, 111; of lesbian,
  33; purified biology and, 107; sameness of, 26, 37, 67, 91, 111, 199, 200,
  275 n. 3; as supplement, 14, 36, 275
  n. 2; white, 14. *See also* Decade of
  the clone; Double; Fantasy; Female:
  body of; Genetic engineering;
  Replication
Clover, Carol, 276 n. 16
Code 46, x, xi, 6, 13–15, 73–74, 78,
  137–76, 280 n. 7, 281 n. 18
Cohan, Steve, 278 n. 10, 279 n. 14, 279
  n. 19
Colebrook, Claire, 283 n. 18

Coleman, Beth, 217

Collage animation, xii, 9, 10, 14, 193, 226–30, 233–56; avant-garde, 231–32

Colonialism, 14, 107–8, 150, 159–60, 244, 252, 275 n. 7; postcolonialism and, 106, 156–58, 172

Come Back to the Five and Dime, Jimmy Dean, Jimmy Dean, 212

Comedy, 6, 274 n. 8

Commodity: culture, 15, fetish, 244, 247, 249

Computer, 62–63, 69, 155, 215, 217, 220; animation and, 189; collage, 284 n. 13; screen, 79, 82–83, 125, 197, 214, 219, 244; software, 71, 191, 196, 219; virus, 73, 210, 213–14

Conceiving Ada, 197, 207

Condit, Celeste, 71, 274 n. 2

Constable, Catherine, 38, 50–52, 56, 64, 276 n. 15

Consumer culture, 68, 77, 80, 86, 228, 266

Contact, 77

Convergence, x, xii, 63, 140, 160, 199, 217, 220, 243, 258, 262, 266, 284 n. 9, 184 n. 3; between digital and genetic, 200, 205, 266, 274 n. 17

Conversion, 205, 208, 213, 224; technical, 222

Cooper, Melinda, 41, 275 n. 6

Copjec, Joan, 124–25

Copy, 2, 9, 28, 38, 84, 96, 100, 111, 136, 216, 251, 265, 267; biological, ix, 194, 265–66; genetic engineering, 128; identical, ix, 102, 262–64; production of, 104; technologies of, 37. *See also* Culture of the copy; Twin

Corporeality: biological, 100; shared, 51, 107

Cosmopolitanism, 155, 157; of cities, xi, 147, 172; cultures and, 139; multiculturalism and, 74

Cousteau, Jacques, 255

Coward, Rosalind, 274 n. 14

Crary, Jonathan, 273 n. 4

Crawford, Joan, 233, 234, 255

Creed, Barbara, 47, 53, 64; on the monstrous feminine, 37, 48–50, 52, 77, 284 n. 8. *See also* Monstrosity

Crick, Francis, discovery of DNA, 69, 71

Crime, xiii, 3, 19, 25, 68, 117–18, 131, 140, 146, 153, 159, 199, 265; criminal intention, 109, 120

Cronenberg, David, 39, 140, 279 n. 21

Cubitt, Sean, 282 n. 1

Cultural imaginary. *See* Imaginary

Culture of the copy, xii, 23, 45, 131, 136, 221, 266. *See also* Schwartz, Hillel

Cut-and-paste techniques, 10, 230, 236, 240, 250; for replicating bodies, 179–94

Cyborg, 38, 42, 44, 51, 58, 61–63, 65, 90, 199, 207, 215, 221, 224, 259

Dadaist, 10, 231, 244

Dali, Salvador, 4

Davis, Betty, 241

Death, 25–32, 41, 48–49, 131, 149–50, 183, 223–24, 256, 260, 269–70, 277 n. 9; drive, 19, 22–24, 26; finality of, 19, 233, 261

Decade of the clone, 11–12, 180

Deception, x, 3, 9, 72, 77, 80, 92, 109–10, 116–17, 120, 122, 129–31, 139–40, 153–54, 208, 218; genetic, 91, 110, 118, 131, 263; sexual, 210–11; spatial, xi; visual, 6, 113

de Certeau, Michel, 146, 159; on oneiric figuration, 145

Deep Blue Sea, 39

Deformity, 38, 44; self-deforming, 154; systems of, 155

Déjà vu, 257–71

De Lauretis, Teresa, 280 n. 2

Deleuze, Gilles, 156–57; on control society, xi, 154–55; on imagined city, 155

Depersonalization, 97, 99, 107

Derrida, Jacques, 161, 265, 281 n. 16; avisual, 264; supplement, 275 n. 2

Detection, xii, 3, 68, 110, 116, 153, 159

Dialogic, 67, 70, 85, 92, 100, 104, 221, 229, 277 n. 1

Dietrich, Marlene, 233–34, 239–41, 245, 253–54

Difference: biological, 25–27, 64, 73, 158; cloning and; 26, 199; racial, 8, 73–74, 160; sameness and, 28, 36, 95, 149, 199–200, 222, 266; sexual, 10, 21, 25–34, 61–63, 105, 118, 136, 149, 163; threat of, 90; visibility of, 15. *See also* Embodiment

Differentiation (non-, un-), 24, 27, 29, 54, 180; of cells, 25–26, 28, 41, 44

Digitization, 70, 179, 181, 185, 214, 236; in cinema, 190–91; in composition, 202; culture and, 8; genetic, 179, 181, 191–92, 194, 205, 266, 274 n. 17; of image, 40, 179–80, 182, 188–89, 192–93; special effects of, 40, 191; techniques of, 15, 180, 187, 190. *See also* Cloning

Disavowal, 9, 45, 89, 106–8, 121

Disease, x, 2, 27–8, 41, 45–46. *See also* Cloning

Disguise, xii, 78, 80, 91–92, 113, 135, 156, 279 n. 14; biological and cultural, x, 84–85, 87, 92; genetic, 188, 114–25, 163; genetic imaginary and, 8, 188; narratives of, 6; sartorial, 77, 86, 91. *See also* Impersonation

Disney, Walt, 229–34, 238–39, 249; Disney World, 226, 245; Walt Disney Company, 233

DNA, xii, 38, 55, 68–71, 81, 86, 155, 196–98, 204, 212, 219, 238, 245, 249,

251, 277 n. 3; alien, 38, 51, 82–83, 87–89, 92, 115, 128–29; double helix, 1, 3, 4, 69, 71, 80, 130; fetish of, 274 n. 12; fingerprinting and, 157; human, 39, 77–78; mixing/hybrid, 38–39, 43, 74, 81–83, 226; molecule, 71; patenting of, 244; recombinant, 77; sequence, 12, 42, 77, 115, 121, 128; testing for, 68, 141, 147. *See also* Genetic engineering

DNA, 39

Doane, Mary Ann, 47, 89, 103, 108–10, 119–21, 273 n. 6, 274 n. 13, 282 n. 2, 282 n. 3

Documentary, 159, 235–36

Dolar, Mladen, 163–64

"Dollie Clones, The," 197. *See also* Hershman-Leeson, Lynn

Dolly the sheep, ix, 4, 11, 27, 181, 192–93, 227, 273 n. 2, 275 n. 8; *See also* Franklin, Sarah

Donald, James, 139, 141, 145–46, 149, 159, 162

Donaldson, Roger, x, 48, 68, 75–76, 79, 83, 86, 115, 152

Doppelgänger, 95, 98–99

Double, 47, 29, 127, 150, 196, 202, 264, 270; clone and, 101, 259. *See also* Agency

Double Indemnity, 196

Double take, 14, 257–71

Douglas, Mary, 284 n. 1

Downloading, 104, 207, 267. *See also* Copy

Duden, Barbara, 273 n. 4, 273 n. 5, 280 n. 6, 283 n. 2

Duplication, 55, 62, 64, 67, 99, 118, 182, 259–60, 266; biological, 55, 98, 265; cloning as, 109–10; genetic, 31, 90, 104, 108; identical, 29; monsters and, 37

Duplicity, xi, 89, 91, 102–4, 109–10, 119, 127, 128, 135–36, 196, 218, 261, 263

Dunaway, Faye, 281 n. 17, 283 n. 13
Dutoit, Ulysse, 11
Dyer, Richard, 104, 132, 198, 274 n. 22, 278 n.10, 282 n. 4, 282 n. 11

Eco, Umberto, on history of expositions, 244
Edelman, Lee: on Baudrillard, 20–22, 30–31; on reproductive futurism, 21
Edgar, Andrew, 274 n. 12
Egg, 41, 226, 236, 238, 241, 242, 246–47, 255–56; donation of, 32–33, 188, 228, 244; fertilization of, 27, 32
Ego, 97, 101, 104, 216; autonomous, 36; ideal, 98–99, 105, 260
Egorova, Yulia, 274 n. 12
Eighteenth Angel, The, 39
Ellis, John, 274 n. 14
Embodiment, ix, 36, 170, 172, 187, 192, 202, 207, 217–18, 221, 228, 263–64, 270, 276 n. 23; cloning and, 37–38, 43, 48, 57, 110, 150, 166–67, 212; colonial, 159; cross-species memory, 53; desire and, 116; difference and, 10, 221, 266; feminine, 14, 48, 90, 227 n. 1; genetic, x, 2, 8, 16, 77, 86, 100–111, 166, 186, 192; masculine perfection of, 129, 279 n. 21; monstrous, 37, 48; panhuman, 157; posthuman, 62, 85; prosthetic, 130; psychic, 99; racialized, 108; sameness and, 55; sexual, 169, 181; shared, 117–18, 130; uncanny and, 175; white beauty in, 86, 133
Embryo, 12, 81, 115, 125, 255; embryology, 241, 276 n. 8; screening of, 68; splitting of, 148; transplantation of, 226
Emotion, 11, 49, 61–62, 75, 80–82, 84, 123, 183, 224. See also Affect
Empathy, 43, 62, 82, 146, 212; empathy virus in Code 46, 149, 155, 159–60, 164–65
Enlightenment, 139, 149, 151, 156, 251

EPCOT Center, 245
Equilibrium, 240
Eternal Sunshine of the Spotless Mind, 282 n. 18
Eugenics, 2, 123, 135; use in the past, 107, 125, 161. See also Aesthetic
Evolution, 25–33, 228; biology and, 24, 30; reverse, 21
Evolution, 55, 73
eXistenZ, 140, 179 n. 21
Expositions, 226, 241–46
Fairground, 226, 228–29, 232–33, 237–38, 240–41
Fanon, Frantz, on psychoanalysis and colonialism, 106–7
Fantasy: authorship and, 126; autopoietic individuality and, 53; of cloning, 91, 96, 98, 134, 263, 265; cultural, 236; of genetics, 43, 45; of immortality, 4, 22; projective, 133; of technology and innovation, 9, 15; reproductive body and, 250
Father, 62, 87, 221, 235; cloning of, 29, 46
Female, 32, 57, 75, 82, 85, 108, 206, 210, 212–3, 215, 236, 240–41, 276 n. 15, 278 n. 12; alien, 74; autogenerative body of, x, 87; body of, 25, 48, 74, 78, 91, 218, 231, 241–42, 249, 252, 261; cloned, 56, 214; diseased, 48; DNA of, 212; genetically engineered, 77; homosocial triangulation of, 13, 133, 279 n. 24; hybrid, x; masculine, 279 n. 15; reproductive body of, 72, 252; sexuality, 77, 199, 214; white, 82. See also Body; Monstrosity: monstrous body
Femininity, 56, 58, 85–87, 90, 101, 103–4, 109, 119–21, 127–28, 133, 196–98, 206–8, 211, 214, 216, 231, 244, 261, 276 n. 15, 277 n. 1, 278 n. 12; artifice and, 13, 85, 136, 214; automated feminine, 80, 90; code/encoding of,

Femininity (*cont.*)
68, 89; commodified, 85; duplicity
of, 104; genetically engineered, 88;
mimicry of, 5; narcissism of, 105,
277 n. 2; white beauty and perfection of, x, 14, 58, 74–75, 82, 86, 90,
98, 173, 196, 200, 215, 240, 249. *See
also* Embodiment; Monstrosity;
Woman

Feminism, ix, xiii, 31–33, 90, 103, 207

Femme fatale, xi, 77, 87, 90–92, 110,
120, 159, 172, 174, 196–97, 210, 212,
218; as monstrous, 89, 92. *See also*
Genetic engineering

Fetishism, xi, 246, 249; of gene, 9;
technological, 46, 89, 119, 186. *See
also* Commodity fetish; Mulvey,
Laura

Fetus, 53, 69, 81, 141, 148, 194; cloning
of, 73–74, 138, 149, 155–56, 163, 171,
173, 193

Fifth Element, The, 73, 140

Film noir, 77, 140, 196–97, 199, 218

Film theory: feminist, 37, 118–19, 128,
136, 261; psychoanalytic, 217 n. 14

Fincher, David, 38

Fletcher, John, 278 n. 11

Fly, The, 38–39, 239

Ford, Harrison, 56, 159

Fordism, 155; production line in, 228,
244, 247

Fortier, Anne-Marie, 280 n. 13; on
inter-ethnic propinquity, 157

Foster, Gwendolyn Audrey, 274 n. 22

Foucault, Michel, 154, 273 n. 4, 273
n. 5, 280 n. 6, 283 n. 2; on Enlightenment, 149, 156; on life itself, 5, 280
n. 6; on surveillance, 137

Frankenheimer, John, 39

Frankenstein, 47

Franklin, Sarah, 37, 69, 82, 118, 163,
186, 192–94, 218, 220, 273 n. 2, 273
n. 5, 274 n. 9, 274 n. 2, 274 n. 11, 274

n. 20, 275 n. 5, 275 n. 8, 275 n. 11,
276 n. 11, 276 n. 22, 277 n. 2, 278
n. 7, 279 n. 16, 280 n. 1, 280 n. 6, 280
n. 12, 280 n. 15, 280 n. 26, 283 n. 2

Fraud: detection of, 153, 159; identity
and, xi, 3, 68, 120, 125–26, 160. *See
also* Impersonation

Freak, 14, 39, 238, 241, 254

Free association, 143, 227–28; technique of, 231

Freud, Sigmund, 145, 280 n. 5; on
narcissism, 101, 104–5; on uncanny,
149, 163, 261

Friedberg, Anne, 277 n. 8

Fuss, Diana, 106–7, 110

GATTACA, x, xi, 1–7, 13–15, 55, 72–74,
113–35, 140, 147, 150, 152–53, 227,
273 n. 1, 278 n. 1, 279 n. 23

Gays, 57, 111, 132, 278 n. 5; male cloning culture and, 276 n. 19. *See also*
Lesbians and gays

Gaze: cinematic, 114–15; medical, 210;
panoptic, 220; scientific, 8, 122

Geiger, H. R., 277 n. 10

Gender: confirmation of, 34; identity
and, 55–56, 127; impersonation, 127,
279 n. 18; language of, 129; parody
of, 11; performative, 110, 279 n. 19;
spectatorship and, 120, 219; terrorism and, 199, 210, 213. *See also*
Femininity; Masculinity

Genealogy, 21, 95, 139, 174, 218,
265–66; kinship and, 116, 130;
globalized, 157; linear, 150

Genetic Admiration, x, xii, 9, 14–15,
225–56, 284 n. 18

Genetic engineering, 82, 210; affective
dimensions of, 35, 182–83; artifice,
227; authenticity and originality
(loss of), xii, 6, 14, 182; biology and,
x, 13, 73, 87, 97, 115, 147, 162, 192;
biosociality and, 74–75; body, xii, 6,
9, 73, 79–80, 91, 137, 147–48, 151–52,

157, 165, 187, 192–94, 220, 223, 250, 256, 265; debates on, 11, 15, 19; desire and, xiii, 10, 228, 233; double/duplication and, 91, 150, 263; fantasy of, 41, 146; femme fatale, 74; of fetus, 193; film and, 12, 140, 150, 239; future and, xiii, 43, 165, 171; horror and, 38, 40; of human, 140; of hybrid female, x; imitation/impersonation and, xi, 2, 7, 68, 95–97, 128–29, 183, 260; incest and, 161; kinship and, 149, 162, 168–70, 172, 194, 242, 280 n. 18; life and, 8, 12; miracle of, 81, 242; monsters and, 37–38, 43, 90, 92, 229; recombinant science of, 255; replication by, 91; of reproductive body, 195; temporalities and, 162–63; visualization of, 64; white body and, 14; wonder of, 240. *See also* Baudrillard, Jean; Bio-aura; Cloning; Copy; Femininity; Oedipal complex

Genetic imaginary, xii, 7–10, 16, 20–21, 32–34, 38, 46, 90, 96, 136, 162, 182, 222, 228, 238, 240, 258, 263–65, 269–71, 274 n. 12, 276 n. 11; bio-aura and, 187–88; clone and, 24, 77, 107, 187–88; condensation and, xiii; desire and, 11, 100, 264; kinship and, 46; landscapes and, 33, 186, 226; legibility and, 68

Genetics. *See* Aesthetic; Artifice; Body; Disguise; Embodiment; Impersonation; Manipulation; Sexuality

Genome, 12, 23, 70, 72, 78–85, 137–40, 146, 151, 155, 180–81, 223, 228, 233, 244, 250–56, 284 n. 3; surveillance and, 74, 156

Genre. *See* Biohorror; Body-horror film; Horror; Melodrama; Natural history film; Romance film; Romantic comedy; Science fiction; Thriller; Western

George, Susan, 125

Ghost, 27, 58, 156, 164, 170–71, 215, 268; traces, 22, 258

Gibson, Pamela Church, 275 n. 4, 276 n. 13

Giddens, Anthony, on disembedding and reembedding, 280 n. 6, 284 n. 4

Gilloch, Graham, 184

Gledhill, Christine, 282 n. 11

God, 4, 63, 186; godlessness, 224; and the godlike, 73, 150; God's-eye view, 141, 143, 145, 175

Godsend, 150

Gonder, Patrick, 71–73, 161

Gondry, Michel, 281 n. 18

Gordon, Andrew, 274 n. 22

Gordon, Douglas, 285 n. 4

Govil, Nitin, 280 n. 7

Gromala, Diane, 152; on remediation, 282 n. 12

Grosz, Elizabeth: on architecture, 138; on mirror stage, 97–98; on Möbius strip, 273 n. 3; on monsters, 37

Grotesque, 39, 48; in teratology, 43, 45, 276. *See also* Body

Grusin, Richard A., 282 n. 12

Gu, Felicity, 146

Guattari, Félix, 155

Gumbrecht, Ulrich, 185, 188, 281 n. 1

Gunew, Sneja, 156

Halberstam, Judith, 279 n. 15

Hall, Stuart, 274 n. 14

Hamm, Nick, 150

Haptic quality, 278 n. 2

Haran, Joan, 274 n. 15, 274 n. 16, 274 n. 18, 274 n.19, 281 n. 19, 284 n. 3

Haraway, Donna, 186, 273 n. 5, 274 n. 3; on corporealization, 5; on cloning, 46–47; on gene, 5; on gene fetishism, 9; "A Manifesto for Cyborgs," 207; on phantom object, 4

Hark, Ina Rae, 278 n.10

Harlin, Renny, 39

Hart, Lynda, 282 n. 2
Hartfield, John, 231
Harvey, Sylvia, 90
Hawke, Ethan, 2, 3, 114, 122, 134
Hayles, Katherine, 16, 85, 104, 217; on
    computational universe, 72, 74; on
    posthuman, 15, 62–63, 68, 87, 276
    n. 21
Helgenberger, Marg, 74–75, 83
Hemmings, Clare, 283 n. 18
Hennion, Antoine, 185–86
Henriques, Julian, 274 n. 14
Henstridge, Natasha, 74, 76, 85–86
Heroine, 13, 38, 50, 56, 58, 206, 216,
    219, 222, 233, 276 n. 16
Hershman-Leeson, Lynn, x, xii, xvi,
    140, 196–97, 201–2, 206, 208–9,
    211, 213, 216, 220, 227
Heterosexuality, 32–35, 54–57, 59,
    63, 86, 107, 111, 117, 119, 133, 139,
    141, 168, 172, 206–10, 216, 221, 266;
    complementarity and, 211; con-
    ventionality, 212; desire and, 15, 82;
    difference and, 21, 30–31, 118; gene-
    ticized, 216; homosexual distinction
    and, 56, 131, 136; reproduction of, 11,
    22, 27–30, 33, 90, 149–50, 204. See
    also Kinship
Hills, Elizabeth, 275 n. 4
Hills, Matthew, x; on species-level
    biohorror, 38–39, 87, 277 n. 11
Hitchcock, Alfred, 117, 215, 267, 279
    n. 20
Hobson, Dorothy, 274 n. 14
Höch, Hannah, 231–32, 239–40
Hollway, Wendy, 274 n. 14
Hollywood, ix, 56, 196, 198, 200,
    207–8, 210, 222, 240, 253–54, 267;
    cinema, 119, 191, 199, 214, 216–18,
    240, 277 n. 9; dream, 56; feminini-
    ties, 208; machine of, 198, 221; stars,
    206, 214, 216, 226, 228, 279 n. 18

Holmlund, Chris, 276 n. 16, 278 n.10,
    279 n. 14
Homoeroticism, 13, 55–56, 59–63, 65,
    118, 167–68; desire, 55–56; doubling,
    58; dynamic, 59; impersonation,
    118, 134
Homosexuality, 30–31, 118–19, 131;
    cloning and, 57, 118; desire and, 57,
    179 n. 24; narcissism and, 101, 168
    parenting and, 34, 56; prohibition
    against, 169; triangulation and, 13,
    133. See also Heterosexuality; Lesbi-
    ans and gays
Honey, I Shrunk the Kids, 284 n. 7
Horror, 39, 47, 50, 56, 77. See also
    Biohorror; Body-horror film
Hulk, 39
Hulme, Peter, 273 n. 4
Human Body, The, 235–36
Human Genome Project, 12, 181, 234
Hurley, Kelly 275 n. 4
Hybridity, x, 14, 39, 107, 208, 233, 239
Hymans, Peter, 277 n. 11

Icon, 228, 233, 240, 242, 254, 258, 283
    n. 14; cultural, 46. See also under
    Lesbians
Identical twin. See Twin
Identity: dialogic of, 67, 85, 92, 109;
    fabricated, 13, 111–36; legibility of,
    9, 91, 129, 156; mistaken, xii, 6, 13,
    140, 263. See also Impersonation
Illegibility, 11, 74, 77, 87–89. See also
    Legibility
Image: culture of, xii, 80, 180, 188, 269;
    indexicality of, 188–91; mutability
    of, 16, 40–41; pixelated, 196–97,
    214; reproducibility of, 13, 15
Imaginary, 9–12, 20, 80, 97–99, 219,
    229, 234, 236, 251, 258–59, 262;
    abject, xiii; clone, 102, 105, 270; cul-
    tural, x, 15, 96; social, 8; teratologi-
    cal, 41. See also Genetic imaginary

Imitation, 9, 20, 22, 57, 92, 97–102, 121, 128, 151, 194, 232, 260, 262, 279 n. 14; alien, 84; artificial, 27; clone, 95, 107, 183, 195–96, 212, 216–7; cultural, 22, 198; history of, 102; masculine, 119–20; technologies of, 1, 7, 92, 116

Immortality, 5, 183; desire for, 4, 24–26, 233; genetic, 31; unbearable, 29

Impersonation, 15, 67, 72, 85, 106; cloning and, 110; genetic, xi, 2, 6, 11, 13, 87, 96, 113–36; queer, xi, 113–136, 279 n. 18

Impostor, 110, 122–27, 136, 259

Impurity. See Purity

Inanimate in cinema, 196, 226, 260, 268–70

Incest, 18, 160, 170, 211; cloning and, 26, 73; desire and, 27, 276 n. 13; prohibition of, 13, 148, 155, 161, 167, 169; sexual merger and, 51; as transgression, xi, 141, 168. See also Oedipal complex

Incredible Shrinking Man, The, 284 n. 7

Indexicality, 188–91, 281 n. 3. See also Image

Individuality, 6, 10, 15–6, 22, 47–48, 53, 63, 65, 142, 150, 187, 200, 265

Inheritance, 47, 139, 174, 129; biological, 96, 111, 117, 162

Instinct, 59, 62, 78, 80, 101; biological, 13; drive, 82, 86; memory and, 50–51; self-preservation, 97

Intelligibility, xiii, 72, 196, 264; body 121, 222–24; cultural, 68; human, 34; visual, 10

Interface, xii, 40, 217, 283 n. 16; human-machine, 203; human-technical, 181, 195; sexual, 63, 219

Interiority, 123, 138, 166; biological, 6; imaginary, 99, 164

Intersubjectivity, 19, 182; embodiment, 50, 53

In This World, 280 n. 11

Invasion of the Body Snatchers, 47

Invisibility. See Visibility

In vitro fertilization, 32–33, 46, 52–53, 188, 192, 194, 235, 241

Irigaray, Luce, 103, 105, 109, 276 n. 15, 277 n. 1

Island, The, 118, 140, 150

Island of Dr. Moreau, The, 39, 227 n. 11

Jackson, Michael, 57, 276 n. 17

James, Joy, 284 n. 11

Jameson, Fredric, 197

Janet, Pierre, on legendary psychasthenia, 97. See also Caillois, Roger

Jennings, Ros, 276 n. 16

Jeunet, Jean-Pierre, x, 37, 40, 45, 58, 60, 64, 118, 150

Johnston, Claire, 278 n. 11

Johnston, Joe, 284 n. 7

Jordanova, Ludmilla, 273 n. 4

Jurassic Park, 39, 78, 150, 277

Kac, Eduardo, 5, 283 n. 14

Kalin, Tom, 117

Kaplan, E. Ann, 282 n. 2, 282 n. 10

Kavanagh, James, 275 n. 4

Keller, Evelyn Fox, 69, 71, 138, 274 n. 2; on secrets of life, 250–51

Kember, Sarah, 4, 9, 46, 274 n. 3, 280 n. 2

Khanna, Ranjana, 275 n. 7

Kinder, Marsha, 282 n. 7; on intelligent machine, 207

King Kong, 253

Kingsley, Ben, 74–75, 83

Kinship, ix, 7, 10, 30–33, 46, 116, 130, 155–59, 161–63, 168–69, 193–94, 213–14, 217–18, 242, 275 n. 11, 278

Kinship (*cont.*)
n. 5, 278 n.7, 279 n. 22, 282 n. 17;
genetic, 29, 43, 148–49, 160, 164,
170–72, 218, 280 n. 1, 280 n. 15;
heterosexual, 8; improvization and,
113–37; queer, x, 117–18, 130–31,
278 n. 8; theory of, 161, 274 n. 20;
transgenic, 13, 38, 44; warped, 13,
167, 171–72, 266
Kirby, David, 116
Kiriya, Kazuaki, 74, 140
Kirkham, Pat, 278 n. 10
Kracauer, Siegfried, 6
Kristeva, Julia, 277 n. 1; on abjection, 47–48, 284 n. 8; on clean
and proper body, 53
Kuhn, Annette, 273 n. 7, 278 n. 3, 280
n. 3, 282 n. 10

Laboratory, xi, 38, 42–44, 46, 58,
68, 78–82, 84, 107, 122, 153, 179,
226–27, 234, 239, 250, 259, 283 n. 14
Lacan, Jacques, 124–26; on language,
109; on mirror stage, 97–98; on
uncanny, 162–63
Lady from Shanghai, The, 196
Lai, Larissa, 159
Lamar, Hedy, 206
Landecker, Hannah, 6–7, 274 n. 10
Landis, John, 57
Lasseter, Joe, 284 n. 7
Last Time I Saw Paris, The, 206
Latour, Bruno, 185–86, 277 n. 3
Lauritsen, John, 276 n. 19
Law, Jude, 1, 3, 114, 116, 122, 132, 279
n. 21
Lean, David, 147
Le Breton, David, 274 n. 12
Lee, Ang, 39
Leeming, Frances, x, xvi, 9, 226,
230–31, 234, 238–39, 243, 247, 254
Legibility, 72, 87–88, 92, 192, 264;
biological, 89, 156; body, 68, 72–73,
77, 82, 91, 121, 147, 195, 200; genetic,

70–71, 78, 147, 196; nature, 69;
of white femininity, 74. *See also* Genetic imaginary; Identity; Illegibility
Lehman, Peter, 278 n. 10
Lesbians, 128; cloning and, 33; continuum and, 212; as couple, 214;
desire of, 60; in film, 276 n. 20; as
icon, 56–59, 63. *See also* Weaver,
Sigourney
Lesbians and gays: adoption by, 33;
parenting by, 32; partnership of,
33, 275 n. 10; passing of, 279 n. 23;
politics of, 31; sexualities of, 111
Levin, Maud, 231
Lévi-Strauss, Claude, 161, 280 n. 16
Life itself, 5, 9, 28, 64, 78, 81, 92, 125,
147, 216, 229, 276 n. 22. *See also*
Foucault, Michel
Lindee, Susan, 46, 274 n. 2, 276 n. 10,
277 n. 2
Lippit, Akira, 264–65; on the avisual,
264. *See also* Derrida, Jacques
Lorenzo's Oil, 150
Los Angeles, 82, 84–85, 277 n. 10
Low, Gail, 278 n. 4
Lowe, Andrew, 274 n. 14
Lucas, George, 118
Lundin, Michael, 62, 276 n. 21
Lury, Celia, 37, 193, 275 n. 5, 280 n. 26,
280 n. 12, 283 n. 2
Luyken, Gunda, 232
Lykke, Nina, 15, 25

Machine, 44, 51, 71, 199, 203, 217,
220–21, 231; intelligent, 198, 207
Mackenzie, Adrian, 204–5, 211, 215,
274 n. 17; on indeterminacies of
biotechnology, 199, 202–3; on
transduction, xii, 203
Madsen, Michael, 74–75, 83
Malignancy, 45, 55; in disease, 45–46;
growth of, 27–28, 44, 83
Manipulation, 227, 251; biogenetic,
xi; of bodies, xi; cell, 40, 84; digital,

188, 192–94; genetic, 2, 4, 8, 11, 40–41, 43, 115, 181, 194, 196, 264; technical, 181, 183, 224. *See also* Digitization

Manovich, Lev, 189–91, 282 n. 1

Man with the Golden Arm, The, 206, 214, 216

Marchessault, Janine, 273 n. 4, 284 n. 6

Marks, Laura, 278 n. 2

Marks, Ross Kagan, 55

Marrinan, Michael, 185, 188, 281 n. 1

Masculinity, 87, 103–5, 110, 117–19, 122–29, 135, 210, 229, 244, 249, 253, 279 n. 13, 284 n. 10; adventure narratives of, 10, 234, 254; detachment of, 86, 235–36; fantasy of self-regeneration and, 53; femaleness and, 279 n. 15; male redundancy and, 87, 90; masculine body, 56, 279 n. 14; masculine desire, 72, 104, 122, 134, 198, 251; masculine ego, 249, 254; masculine subject, 103–5, 125–26; perfection and, 114, 118, 128, 134, 136, 279 n. 21; reproduction of, 87, 250; spectacle of, 278 n. 10; vanity of, xiii. *See also* Agency; Masquerade; National hero; White(s)

Masquerade, 6, 13, 85, 128, 212, 278 n. 11; masculine, 119–21, 124, 279 n. 13, 279 n. 19; theory of, xi, 112, 118–19, 282 n. 3; womanliness (or femininity) as, 103, 108–10, 119–21, 136, 196, 278 n. 12. *See also* Doane, Mary Ann; Riviere, Joan

Materiality, 15, 187, 213, 217

Maternal body, 48, 53, 125, 170–71, 261, 269

Matrix: heterosexual, 30, 188; universal, 221

McKinnon, Susan, 118, 163, 194, 218, 274 n. 20, 275 n. 11, 278 n. 7, 280 n. 1, 280 n. 15

McNeil, Maureen, 235, 245–47, 281 n. 23, 284 n. 10

Medak, Peter, 277 n. 11

Méliès, Georges, 243

Melodrama, 226, 233, 240, 255

Memento, 281 n. 18

Memory, 50, 141, 164–66, 170, 173–74, 255, 261, 276 n. 14; commoditization and, 276 n. 14; cross-species, 51–53; dislocation and, 62; genetic kinship and, 43; memory of religion in art, 185, 194; prosthetic, 141, 155, 215, 281 n. 18

Mercer, Kobena, 276 n. 17

Merck, Mandy, 276 n. 20

Metastasis, 26–28, 204, 221

Microbiology, 204, 282 n. 9

Microinjection, 4, 81–83, 87, 92

Microscope, 235, 237, 244, 250–51

Migrants, 142, 156–57

Miller, George, 150

Mimic, 277 n. 11

Mimic 2, 277 n. 11

Mimicry, 2, 64, 95–99, 102–8, 110, 207, 219, 254–55, 263; in art, 194; bio-aura and, 15; biogenetic, 121; colonial/postcolonial, xi, 106–8; comic, xiii; double, 254–55, 274 n. 4; feminine, 57, 105; gender and, 34; morphology of, 97; racialized scenarios and, 14; savagery and, 274 n. 4; theory of, 13. *See also* Biomimicry

Minghella, Anthony, 279 n. 21

Minority Report, 278 n. 3

Mirror stage, 97–99

Miscegenation, 43, 73, 107, 160

Mise-en-scène, x-xii, 1, 4, 6, 8, 20, 53, 60, 117, 140–41, 143, 147, 152–53, 166, 170, 175–76, 196, 198–99, 208, 218–19, 227–28, 283 n. 14

Misrecognition, xii, 46, 67, 78, 99, 148, 152, 172, 220, 255, 263, 266

Mistaken identity, xii, 6,13, 140, 263.
    See also Identity
Mitchell, W. J. T., 4–5, 12, 179–82, 188,
    192–93, 281 n. 2, 284 n. 13
Modes of perception, xii, 180–81, 190,
    194, 251
Modleski, Tania, 279 n. 20, 282 n. 10
Molina, Alfred, 75, 83
Monster, 14, 52, 76, 81–82, 241, 259;
    as alien, 38; duplication and, 37;
    genetic engineering of, 36, 42–43,
    73, 90; as marvel, 40–41, 229, 241,
    259, 283 n. 1, 284 n. 2; maternal, xiii,
    53; of nature, 46; as sibling, 50. *See
    also* Genetic engineering
Monstrosity, 37, 50–51, 74, 88, 207,
    241; as alien, 38, 83; birth of, x, 14,
    44, 48, 241; cloning and, 41–42,
    46–49; double and, 47, 280 n. 10;
    maternity and, xiii, 56; monstrous
    body, 39–41, 45–46, 49, 77, 85, 275
    n. 4, 276 n. 7, 277 n. 10; monstrous
    child, 52; monstrous feminine, xi,
    37, 48–49, 74–75, 77, 87–92, 284
    n. 8; in siblings and predecessors, 55
Morphing, 40, 50–51, 53, 188
Morris, Meagan, 141
Mortality, 25, 27, 29, 95. *See also*
    Immortality
Morton, Samantha, 140, 153,
    168–69, 174
Moseley, Rachel, 282 n. 11
Mother, 91, 98, 116, 125, 164, 169–72,
    193, 221, 282–3 n. 13; bad, 51, 53;
    child reunion and, 52; cloning
    without, 29, 46, 56; good, 51, 53;
    regressive (mother) love, 60; sur-
    rogate, 50, 53
Mourning and melancholia, 15, 19, 187
Multiculturalism, xi, 14, 74, 156–57,
    172; diversity, 159; flows, xi; inti-
    macy and, 280 n. 13
Multiplicity, 55, 117, 150, 275 n. 1

Multiraciality, 42, 135, 153
Mulvey, Laura, 259–61, 268, 277 n. 12,
    282 n. 2, 282 n. 10, 285 n. 5; on de-
    layed cinema, 267–68; on entranc-
    ing illusions, 270; on visual pleasure
    of narrative cinema, 122, 168
Mutational bodies, 38–40, 42, 51, 74,
    76, 80, 88, 91, 215

Narcissism, 60, 101, 105, 167, 277 n. 2;
    desire and, 57, 167; homoeroticism
    and, 55–56; homosexuality and, 57,
    168; primary and secondary, 101;
    recognition of, 60; techno-scientific,
    101. *See also* Freud, Sigmund
Narrative: agency in, 251; closure in,
    49–50, 52, 122, 165, 174, 254; drive of,
    49, 77, 165, 168; progression of, 251
National hero, 227, 234, 246, 249, 254,
    284 n. 10
Natural history film, 226
Neale, Steve, 278 n. 10
Nelkin, Dorothy, 4–5, 46, 274 n. 2,
    276 n. 10, 277 n. 2
Neumann, Kurt, 38, 239
Newman, Kim, 273 n. 7
Newton, Esther, on drag, 127
Newton, Judith, 275 n. 4
Niccol, Andrew, x, 1, 3, 7, 113–14, 117,
    124, 132, 134, 140, 227
Nicholson, Jack, 283 n. 13
1984, 281 n. 21
Nolan, Christopher, 281 n. 18
Nondifferentiation, 24, 26
Nonhuman, 5, 38–39, 51–52, 61–62,
    64, 199, 202, 207–8, 221–22, 252, 255
Nottingham, Stephen, 274 n. 19
Novak, Kim, 206, 214–16
Now, Voyager, 241
Nyman, Michael, 113

Oedipal complex, 170–72, 215, 267; as
    foundation of culture, 19; genetic
    engineering and, 140, 163; incestu-

ous transgression and, xi, 160–61, 169, 173; kinship and, 162; non-Oedipal structures and, 149; post-Oedipal reproduction and, 13, 266, 269; romance and, 166; sexuality and, 29; transgression in, 264

O'Keefe, John, 199

Omnipotence, 123, 141, 143, 145, 175, 268; desire for, 164; fantasy of, 46, 106, 139, 157, 250

Ontogenesis, 204, 215

Ontology, 86, 175, 204, 223, 259, 265; insecurities of, 165, 267–69

Orbaugh, Sharalyn, on techno-orientalism, 148, 280 n. 8

Organism, 12, 25, 63, 97, 180, 195

O'Riordan, Kate, 142, 281 n. 19

Ostherr, Kirsten, 5, 281 n. 3

Otherness, 24, 28, 37, 51, 98, 103–5, 108, 200, 263, 275 n. 7

Out of the Past, 147

Panhumanity, 135, 280 n. 26, 280 n. 12

Pastiche, xii-iii, 104, 197–99, 218, 282 n. 4

Passing, 38, 49, 61, 90, 110, 118, 121, 130–31, 220, 278 n. 5, 279 n. 23; as human, 57; racialized, 279 n. 23

Password, 72, 155–56

Paternity, 63, 87–88, 126, 279 n. 17; authority of, 136

Patriarchal systems, 48, 88, 103, 119, 120, 136, 207; culture and, 103, 108

Pattillo, Alan, 230

Pattison, Stephen, 274 n. 12

Pearce, Lynne, 275 n. 5, 280 n. 4; on dialogic theory, 277 n. 1

Penley, Constance, 273 n. 7, 275 n. 4, 278 n. 3, 284 n. 6

Perceptual disturbance, 182, 187, 192, 257, 266

Perlman, Ron, 65

Phallus, 59, 103, 130, 159

Phantasm, 19, 23, 80, 96, 260–62; of city, 139, 162–63; genetic project and, 20–21, 33; phantasmogoria and, 162, 188, 228, 245

Phenomenology, 102, 264, 283 n. 17

Phillips, Adam, 66–67, 91, 262–63, 274 n. 1

Photography, 4, 99–102, 121–22, 189–90, 192, 221, 223, 232–33, 235, 259, 267; authenticity of, 190; reproduction in, 5, 181, 188–89; technique of, 180, 189, 229, 284 n. 11, 285 n. 5. *See also* Barthes, Roland

Pinon, Dominic, 65

Pixel, 196–97, 275 n. 5

Polanski, Roman, 281 n. 17, 282 n. 13

Pollack, Robert, 88–89

Popular culture, 10, 46, 51, 61, 85, 148, 229, 252

Postcolonialism, xi, 106, 156–58, 172

Posthuman, 15, 62–63, 65, 68, 90, 111, 207, 276 n. 21; culture, 87; cyborg as, 90; embodiment of, 62, 85, 276 n. 21; life-form, 8; media convergence amd, xii; mutability of, 63, 75; reproduction and, 170

Powers of Horror, 47

Pregnancy, 141, 187, 221, 248–49, 254, 276 n. 7, 276, n. 23; alien, 48, 53, 64; body during, 149, 252

Preminger, Otto, 206, 216

Procreation, 29–30, 46, 107, 188, 249

Progenitors, 77, 88, 193, 199

Prosthesis, 23, 82, 141, 208; intimacy of, 131; memory and, 155, 164–65, 170, 215, 281 n. 18; seduction and, 208; spectacle of, 120. *See also* Embodiment

Psycho, 285 n. 4

Psychoanalysis, xii, 9, 47, 49–50, 52, 56, 66, 108, 159, 182, 264–66, 274 n. 13, 275 n. 7, 275 n. 10, 276–77 n. 23, 278 n. 11; cinema and, 109,

Psychoanalysis (*cont.*)
260; cloning and, 19, 98, 104–5, 262;
colonialism and, 106–7; imaginary
and, 10; masquerade and, 119, 279
n. 18; mistaken identity and, xiii;
sexual difference and, 105. *See also*
Freud, Sigmund; Riviere, Joan

Purity, 107, 275 n. 10; biological, xi,
73–74; vs. impurity, 14, 74, 160,
172–73, 268; racial, 14, 159–61, 187.
See also Femininity; White(s)

Queer: biological processes and, 11,
19; desire, 34, 38, 55–56, 118, 134;
impersonation and, 128; kinship
and, x, 13, 117–18, 130–34, 278 n. 5,
178 n. 8; queering of the social, 13;
sameness and, 21; theory, ix, 136.
See also Feminist theory

Rabinow, Paul, 4, 75; on biosociality, 73

Racializing, 8, 90, 135, 150, 156, 160,
200; purity and, 14, 108. *See also*
Body; Embodiment

Ramis, Harold, 117, 150, 275 n.1

Rapper, Irving, 241

Reassemblage, xiii, 44, 85, 194, 232,
240, 256

Recombination, 139, 194, 219, 228,
232, 238

Recontextualization, 193, 218–9, 249,
255; of body, 217, 231

Reilly, Philip R., 4

Reitman, Ivan, 55

Relic, The, 277 n. 11

Religion, 62, 184–85, 194

Renaturalization, 29, 31, 33, 192

Replicant, 36, 42, 259, 275 n. 2, 276
n. 12

Replication, xi, 9, 31, 63, 96, 179, 182,
200, 222, 224, 275 n. 1, 275 n. 3, 278
n. 4; automatons and, 22; by cloning, 23, 25, 49, 107, 118, 136; self-

replication, 25, 64, 198, 255; sexual,
188. *See also* Genetic engineering

Repression, 108, 125, 149, 162, 166,
176; return of the repressed, 156,
170–71, 261

Reproductive futurism, 8, 21–22

Rich, Adrienne, 211

Ritter, Henning, 184

Riviere, Joan, on womanliness as masquerade, 108, 119–21, 278 n. 12

Robbins, Tim, 142, 153, 168, 175

Roberts, Celia, 273 n. 1

Robocop, 281 n. 18

Robot, 42, 61–63, 90, 140, 224; double
and, 55

Romance film, 140

Romantic comedy, 6

Romantic love, 15, 141, 162, 216

Roof, Judith, xvi, 6, 71, 87, 88, 157, 237,
251, 252, 274 n. 8, 274 n. 2, 274 n. 3,
277 n. 4

Rope, 177

Rose, Nikolas S., 273 n. 5

Rosetta Stone, 89, 282 n. 6

Rothman, Barbara Katz, 274 n. 2

Ryder, Winona, 42, 57, 60, 64

Sacred, 150, 185; sacredness of
life, 179

Said, Edward, 148

Sameness, 28, 32, 33, 36, 38, 62, 67,
90–92, 95, 111, 133, 149, 199, 266;
biological, 26, 55; deathly proliferation of, 21, 29, 49; desire for, 61;
excessive, x, 37, 46, 55, 64; nostalgic
longing for, 26; paradigm of, 30, 35;
repetition of, 60; sexual, 31, 34, 55,
136; transparency of, 90. *See also*
Cloning

Same-sex, 56; civil union of, 275 n. 10;
desire, 60

Sassen, Saskia, 146

Saunders, Desmond, 230

Sawchuk, Kim, 236–38, 273 n. 4, 284 n. 6

Scale: gigantic, 6, 113, 229, 237–38, 242, 248–50, 252, 254; microscopic, 6, 114–15, 121, 124, 157, 226, 232, 236–37, 242, 250, 284 n. 11; miniature, 237–39, 242, 248–50. *See also* Stewart, Susan

Schmidt, Siegfried, 191

Schor, Naomi, 103

Schwartz, Hillel, 122

Science fiction, xi-xii, 6, 13, 20, 37, 46–47, 96, 118, 139–40, 147–48, 150, 158–59, 191, 196–97, 199, 239, 259, 176, 273 n. 7, n. 14, 276 n. 18, 278 n. 3, 278 n. 4, 278 n. 9, 279 n. 17

Scott, Ridley, 36, 38, 77, 140, 278 n. 4

Screened Out, 19, 21, 195. *See also* Baudrillard, Jean

Screening technology, 11, 72, 78, 115, 137

Security, 79, 123, 110, 141, 143, 156–58, 167, 210, 261; ontological, 165; technological, xi, 74, 171

Sedgwick, Eve, 133, 136, 279 n. 24; on relations of the closet, 131

Seduction, 59, 185, 196, 199, 206, 210–13, 216, 219, 271, 283 n. 14; celluloid, 208; prosthetic sexual, 208; sadomasochistic, 58; sexual, 13, 84, 86

Segonzac, Jean de, 277 n. 11

Self: destruction of, 21, 24, 26, 45, 49; duplication of, 15, 100; generation of, 52, 252; image of, 98, 102, 105; insemination of, 32, 188; other and, 98, 99, 101, 260; as same, 104, 107; self-referentiality, 2, 53, 65; self-replication, 25, 64, 198, 255

Sexuality, 19–20, 168, 198, 207, 212, 235; authenticity of, 15, 133; categories of, 21; clone of, 188, 262; control of, 158; copy of, 131; deviance and, 28, 63; difference and, x, 21–22, 27–34, 61, 63, 105, 109, 136, 149, 163; drive, 210; female, 58–59, 77, 86, 128, 199, 214; genetic, 13, 155, 180; identity and, 51, 55, 61, 125; incest and, 161; instrumentality of, 90; interface of, 63, 219; revolution in, 31–33; sameness and, x, 29, 31, 34, 37, 55; theory of, 109, 273 n. 3, 176 n. 18. *See also* Heterosexuality; Homosexuality; Lesbians; Lesbians and gays

Shanghai, 6, 22, 137, 141–48, 152–53, 157–58, 164

Shiff, Richard, 183–84

Shingler, Martin, 279 n. 18

Shock, 14, 83, 180–82, 187, 190, 269

Shohat, Ella, 78, 273 n. 4, 282 n. 5

Siegel, Don, 47

Silverman, Kaja, 276 n. 9, 276 n. 12, 285 n. 5; on double mimesis, 278 n. 4; on excorporation, 261–62

Simondon, Gilbert, 203–4, 211, 213

Simulation, 19, 22, 63, 104, 179, 194

Sinatra, Frank, 214

Singularity, 64, 146, 150, 152, 162; authenticity of, 10, 23, 26, 72, 162, 182; cloning of, 96, 99–100, 105, 107, 111, 150; genetic, 16; of human (body), 49, 72, 95, 179, 182; impersonation and, 133–34, 136; masculine, 117, 126–27, 129, 134, 136

6th Day, The, 55–56, 72, 118, 150

Smelik, Anneke, 15, 25

Sobchack, Vivian C., 273 n. 7, 178 n. 3, 280 n. 3

Sontag, Susan, 27

Space Age, 244–45. *See also* Disney, Walt

Spatiality, 5–6, 98–99, 175–76, 193, 232, 236, 249, 256; relations of, 107, 138, 193; transformations in, 237, 242

Species, 21, 24, 26–28, 31, 38, 48, 51, 56, 62, 180, 193, 195, 244; alien, 48, 63; cross-species or hybrid, 14, 53, 107, 208, 255; degree xerox of the, 22; transgenic, 77, 80. *See also* Biohorror; Metastasis

Species, x-xi, 13–4, 39, 48, 66–92, 115, 152, 277 n. 11

Species 2, 277 n. 11

Spectator, 83, 109, 121, 123–24, 141, 145, 152–53, 164, 166, 169–70, 174, 191, 199, 202, 219–20, 240, 251, 253, 255, 261, 266–68; desire of, 165; identification and, 36; ontology of, 259; pleasure of, 50, 212

Spencer, Christopher, 235

Sperm, 27, 32, 41, 90, 198, 210, 212, 217, 235, 241, 246; donation, 32, 188, 210, 228, 244, 246–47

Spielberg, Steven, 39, 140, 150, 190, 278 n. 3, 279 n. 21

Spielmann, Yvonne, 190, 192–94

Spottiswoode, Roger, 55, 118, 150

Springer, Claudia, 51, 61; on erotic interfacing, 63; on sexual interface 283 n. 16

Squier, Susan, 275 n. 3

Stacey, Jackie, 27–29, 37, 40, 193, 275 n. 5, 276 n. 7, 280 n. 26, 280 n. 12, 283 n. 2; on monsters and marvels, 41, 241, 283 n. 1, 284 n. 2

Stafford, Barbara, 273, n. 4

Star Wars: Episode II–Attack of the Clone, 118

State, xii, 140–41, 146–47, 169, 176

Steinberg, Deborah Lynn, 284 n. 12

Stem cell, 11, 33, 226. *See also* Cell

Stewart, James, 215

Stewart, Susan: on the child, 248; on daydream of microscope, 250; on grotesque body, 241; on miniature and gigantic, 237–38, 248, 250. *See also* Scale

Stingray, 230

Stone, Allucquère Rosanne, 282 n. 7

Straayer, Chris, 279 n. 21, 279 n. 24

Strathausen, Carsten, 156, 164–65, 170; on uncanny, 137, 149–50, 163

Studlar, Gaylyn, 279 n. 13

Subjectivity, 9, 23, 51, 98, 102, 123, 164–65, 170, 261, 264–65; cloning of, 105; colonial, 106; computational, 87; embodied, 283 n. 17; feminine, 276 n. 15; posthuman, 276 n. 21

Supplement. *See* Battaglia, Debbora; Derrida, Jacques

Surrealism, 10, 238

Surrogacy, 9, 32, 53, 188, 235

Surveillance, 72, 122, 137, 148, 153–58, 165, 171, 208, 210, 278 n.3; architectures of, 79–80, 140, 147; genetic, 139, 151; genomic, 74, 140, 156; techniques of, 78, 84; technologies of, 72, 146, 157, 166

Swift, Jonathan, 284 n. 7

Swinton, Tilda, 196–98, 200–202, 206, 209, 213, 215–16, 218, 282 n. 8

Swoon, 117

Symbolic order, 9, 19, 20, 23, 30–34

Symmetry, 2, 59, 123, 183, 218; repetition and, 1, 6, 144, 153–54

Synthetic origin, 58, 61

Talented Mr. Ripley, The, 279 n. 21, 279 n. 24

Tang, Zilai, 146

Tasker, Yvonne, 276 n. 16, 278 n. 10

Taylor, Elizabeth, 206

Teknolust, xi, 13–5, 55–56, 73, 140, 195–224, 227, 255

Television, 80–82, 84, 152, 165, 230–31, 236, 244

Telotte, J. P., 47, 150, 273 n. 7, 278 n. 3, 280 n. 3, 280 n. 10

Temporality, 50, 124, 172, 176; scrambled, 162, 164, 170, 172, 188, 266; warped, 139, 162, 169

Teratology, 46, 241, 275 n. 6, 276 n. 8; imagination of, 37, 41, 284 n. 8; grotesque in, 43, 45; monstrosity of, 49; spectacle of, 55

Teratoma, 41, 44, 46, 65, 241, 275 n. 6, 276 n. 7

Test tube, 50, 227, 235, 244

Thacker, Eugene, xii, 14, 183, 186–87, 217, 274 n. 17, 282 n. 12, 283 n. 17

Theme park, 228, 238, 245

Thompson, Michael, 4

Thornham, Sue, 282 n. 10

Thriller, ix, xi, 6, 80

Thumin, Janet, 278 n. 10

Thunderbirds, 230

Thurman, Uma, 120, 132, 134

Toro, Guillermo del, 277 n. 11

Tourneur, Jacques, 147

Toy Story, 284 n. 7

Transduction, xii, 69, 202–5, 208, 210–11, 213, 215, 217, 222, 224, 282 n. 9

Transnationality, 139; biosecurity and, 155; corporation and, 137; movement and, 139, 143, 280 n. 11

Transparency, 23, 87, 88, 115, 145, 159, 162, 171, 249; biological, 89–91, 158; desire for, 78, 89–90, 137–39, 141, 146, 148, 156, 173, 175–76; diegetic promise and, xi; embodied identity and, 151; Foucault on, 149; genetic (and genetically engineered), 24, 78, 91–92, 121, 129, 146, 151, 163, 167; modernity and, 137–39, 146; myth of, 152; screened body and, 92; of social differences, 75; space and, 143, 151, 153, 156, 175; technology and, xi, 165; visual, 137–38, 143, 148, 153, 163. See also Body

Transsexuality, 212, 282 n. 7

Treichler, Paula, 274 n. 2, 276 n. 10, 277 n. 2, 284 n. 6

Triplets, 14, 56, 197, 199, 244; cloned, 212, 218

Tromble, Meredith, 198

Turney, Jon, 274 n. 2, 276 n. 10, 277 n. 2

Twilight of the Golds, The, 55

Twin, 66, 101; fetal, 172; identical, 95, 99, 101, 259; twinning, 2, 6, 19. See also Doppelgänger

Tyler, Carole-Ann, 98, 110, 277 n. 2

Tyler, Imogen, 176–77 n. 23

Uncanny, 19, 39, 149–51, 160–75; haunting and, 160, 162, 164, 176; identical twin and, 99; resemblance and, 49. See also Strathausen, Carsten; Vidler, Anthony

Unconscious, 11, 20, 26, 91, 100, 106, 135, 186, 257, 263, 264; drive and, 167; fantasies of, 260; fears and desires of, 19, 96, 249, 262; motivation and, 10; processes of, 20, 262; subject formation and, 36; wish fulfillment and, 166

Unheimlich, 149, 261

Uniqueness, 6, 182, 184, 207; of human life, 19

Unoriginal, 111–12, 267

Urbaniak, James, 199

van Dijck, José, 12, 37, 46, 69–70, 88–89, 137; 273 n. 4, 274 n. 2, 274 n. 9, 274 n. 11, 276 n. 10, 277 n. 2, 277 n. 3, 284 n. 6, 284 n. 3; on transparent bodies, 138, 151

Varmus, Harold, 45

Venn, Couze, 274 n. 14

Vera, Hernán, 274 n. 22

Verhoeven, Paul, 281 n. 18

Vertigo, 215, 279 n. 20

Vidler, Anthony, 166, 173; on phobic space, 150; on spatial relations of modernity, 13; on uncanny, 149–51; on warped space, 150

Virtual, 12, 23, 125, 150, 180, 200, 205, 219–21; biovirtual, 140; in cinema, 191; life-forms, 217; reality, 22, 180, 233–34; worlds, 191

Virus, 14, 141, 148, 196, 202, 204, 211, 282 n. 9; computer, 73, 210, 213–14; empathy, 149, 155, 159–60, 164–65

Visibility, ix, 8, 75, 137–38, 156, 172, 264, 267, 281 n. 3; difference and, 6, 15; evidence of, 158, 264; form of, 111; genetics and, 5, 263; vs. invisibility, 5, 10, 263, 280 n. 11; legibility and, 192, 264; literacy and, 78, 88, 90; media and, 12; mental, 164; of racialized bodies, 73, 160; sameness and, 37

Visuality, 264; vs. avisuality, 264–65

Vital Illusion, The, 21

Vitality, 5, 7, 183, 221–22, 224, 258; of human life, 195; imaginary and, 12

Waldby, Catherine, 284 n. 6

Walkerdine, V., 274 n. 14

Watson, James, discovery of DNA, 69, 71, 277 n. 3

Weaver, Sigourney, 37–38, 45, 60, 63; as lesbian icon, 56–57, 59

Weber, Cynthia, 277 n. 7

Weibel, Peter, 195, 277 n. 3

Weimar, Klaus, 183–84

Weinbaum, Alys Eve, 265, 281 n. 1

Weinberg, Robert A., 43

Welles, Orson, 196

West, Paula, 223

Western, 226, 240, 255

Weston, Kath, 130–31, 278 n. 5. *See also* Kinship; Queer

Whale, James, 47

Whitaker, Forest, 75, 83

White(s): adventure narratives of, 14, 108; colonialism, 106; culture, 107, 145–46; desire, 172, 244; detective, 159; fantasy of, 106, 159; and ideal of beauty, x, 14, 58, 74–75, 82, 86, 90, 98, 173, 196, 200, 215, 240, 249; lack of originality and authenticity, 14; purity of, 160; white male, 73, 90, 116, 133, 142, 159–60, 138; white slavery, 42–43. *See also* Body; Cloning; Female: body of; Genetic engineering

Wiegman, Robyn, 34

Wilder, Bill, 196

Willemen, Paul, 190–92, 282 n. 1

Williams, Michelle, 74

Williams, Raymond, 185, 190

Willis, Paul, 274 n. 14

Wimmer, Kurt, 140

Winston, Lord Robert, 235–36

Winterbottom, Michael, x, 6, 137, 142, 144, 153–54, 168–69, 174–75, 227, 280 n. 7, 280 n. 11

Wise, Robert, 77

Wolfram, Eddie, 227, 231, 240

Wollen, Peter, 138, 172–73

Woman, 56, 74, 77, 104, 133, 173, 210, 233, 240, 249, 253, 276 n. 7, 278 n. 11; figure of, 78, 120, 196; image of, 85, 109–10, 119, 136, 215–16, 218; as object, 198; womanliness as masquerade, 103, 108–10, 119–21, 136, 196, 278 n. 12. *See also* Female: body of; Femininity; Pregnancy

Wood, Aylish, 273, n. 7, 278 n. 3

"Work of Art in the Age of Mechanical Reproduction, The" (Benjamin), 1880–88, 280 n. 1

Wright, Frank Lloyd, 124

X-ray, 237, 264

Y chromosome, 198

Young, Lola, 279 n. 23

Zemeckis, Robert, 77

Zombies, 39

Earlier versions of chapters 2 and 5 were originally published in slightly different forms as, respectively, "'She Is Not Herself': The Deviant Relations of *Alien Resurrection*," *Screen* 44, no. 3 (2003), 251–76, and "Masculinity, Masquerade and Genetic Disguise in *Gattaca*'s Double Vision," *Signs* 30, no. 3 (2005), 1851–79, © 2005 by the University of Chicago Press. Some parts of chapters 3 and 5 were reworked for publication in "Screening the Gene: Hollywood Cinema and the Genetic Imaginary," in *Bits of Life: Feminism at the Intersections of Media, Bioscience, and Technology,* edited by Anneke Smelik and Nina Lykke (Seattle: University of Washington Press, 2008), 94–112.

JACKIE STACEY

is a professor of media and cultural studies

at the University of Manchester.

Library of Congress Cataloging-in-Publication Data
Stacey, Jackie.
The cinematic life of the gene / Jackie Stacey.
p. cm.
Includes bibliographical references and index.
ISBN 978-0-8223-4494-0 (cloth : alk. paper)
ISBN 978-0-8223-4507-7 (pbk : alk. paper)
1. Science fiction films–Social aspects.
2. Genetic engineering in motion pictures.
3. Science fiction films–History and criticism.
4. Mass media and culture. I. Title.
PN1995.9.S26S725 2010
791.43'656–dc22 2009042054